SpringerBriefs in Law

More information about this series at http://www.springer.com/series/10164

Ruwantissa Abeyratne

Regulation of Commercial Space Transport

The Astrocizing of ICAO

 Springer

Ruwantissa Abeyratne
Global Aviation Consultancies Inc.
Cote Saint Luc, QC
Canada

ISSN 2192-855X   ISSN 2192-8568   (electronic)
ISBN 978-3-319-12924-2   ISBN 978-3-319-12925-9   (eBook)
DOI 10.1007/978-3-319-12925-9

Library of Congress Control Number: 2014954603

Springer Cham Heidelberg New York Dordrecht London

Springer is part of Springer Science+Business Media (www.springer.com)

Preface

On 4 September 2014, Raymond Benjamin, Secretary General of the International Civil Aviation Organization (ICAO) wrote to Commander Chris Hadfield, former Commander of the International Space Station, inviting him to get involved in a joint United Nations/lCAO initiative designed to plan and promote the safe and efficient development of commercial space flight. The Secretary General in his letter advised Commander Hadfield that in this context, ICAO will host a ground-breaking *UN/ICAO* Commercial Aerospace Symposium from 18 to 20 March 2015 at its world Headquarters in Montreal, stating further that the Symposium will focus on the technological, political, economic and social aspects of commercial space development that are fundamental to safe and efficient commercial space flight.

Typically, and not surprisingly, there was no mention of the most fundamental consideration—legal aspects of commercial space transportation—which would be the cornerstone of any discussion pertaining to commercial space flight. One of ICAO's key functions is in the legal sphere of aviation both in treaty making and in acting as advisor on matters of air law to its member States. Yet, it has shown a feckless insouciance in delivering in this area, and its ineptitude has led to recent legal instruments adopted under ICAO's auspices and guidance—such as the 2010 Beijing Convention and the 2014 Protocol to the Tokyo Convention of 1963—being fundamentally flawed. It also begs the question as to how ICAO could address *inter alia* political and social issues related to commercial space transport without inquiring into the legal possibilities of the Organization's involvement.

ICAO addresses issues of international civil aviation under the Convention on International Civil Aviation (Chicago Convention)[1] and the Council of ICAO adopts Annexes to the Convention exclusively on subjects that pertain to

---

[1] Convention on International Civil Aviation signed at Chicago on 7 December 1944. See ICAO Doc 7300/9: 2006.

international civil aviation. Also, ICAO does not have the structure to sustain an entirely different regime of transport. It barely survives under a meagre budget and is compelled to cut down on existing aviation work programmes. If it is the intent to bring in the regulation of commercial space transportation under ICAO (perhaps by renaming it the International Aerospace Organization) through a separate multilateral treaty; have separate funding and provide the relevant expertise, it might work, as the ICAO model has worked for aviation, and there is no reason that it would not work for commercial space transport regulation. As already mentioned, this approach seminally requires an inquiry into the legal issues involved.

The proposal by ICAO to convene the seminar may have been influenced by a book *The Need for an Integrated Regulatory Regime for Aviation and Space— ICAO for Space?* (Springer: 2011) co-edited by two professors of McGill University and a space scientist from The Netherlands. The book recognizes ICAO as a legislative body, which it is not. It also states that commercial space transportation can be brought into the Chicago Convention by amending the instrument. This is similar to saying that rail transport can be brought into the United Nations Convention on the Law of the Sea (UNCLOS) by just amending the Convention. The 19 Annexes to the Chicago Convention are entirely on civil aviation and there is no practical way in which they can be amended, or added on to or revised, or new Annexes adopted under the Chicago Convention (which is entirely and exclusively on civil aviation) to cover such areas as licencing of spaceports, human space flight, space traffic management, safety of personnel and astronauts and security.

The book then goes on to suggest *inter alia* that an international Space Traffic Management (STM) Organization must be established primarily for the civil and commercial use of outer space. This suggestion, while obfuscating the core claim that commercial space transportation should be regulated under ICAO, is diametrically opposed to a subsequent statement in the book which claims that in order to implement the above-suggested ICAO for space regulatory steps and mechanisms, it may be appropriate to follow an orderly and systematic approach. It bounces back and forth between the STM and ICAO concepts only to conclude that the ideal solution to accommodate space traffic management and other outer space safety requirements would be to amend the Chicago Convention thereby expressly extending ICAO's jurisdiction over space transportation.

If one were to make some sense of this dichotomy, the flirtation with two international organizations would have to be clarified. The most startling recommendation, however, is that ICAO could promulgate a new Annex on "Space Standards". The Study claims that there is precedent for this as well and cites Article 37 of the Chicago Convention which vests in ICAO the authority to promulgate Standards and Recommended Practices as Annexes to the Convention. Of course, Article 37 gives authority, through the ICAO Council to: "secure the highest practicable degree of uniformity in regulations, standards, procedures and organization in relation to aircraft, personnel, airways and auxiliary services in all matters in which such uniformity will facilitate and improve air navigation". There is no mention of spacecraft, space objects, navigation in space or aerospace vehicles in this provision. In the face of this explicit provision, one wonders how ICAO could, without amending Article 37, and revising Article 44 which reflects the aims and objectives of ICAO, which exclusively pertain to aviation, adopt an Annex 20 (in addition to the 19 Annexes on air transport) on space standards.

This book inquires into the legal issues that may be relevant in "astrocizing" (a word coined by the author) ICAO into the commercial space transport regime. Hopefully, it conveys the fundamental message that, what needs to be done by ICAO at the outset is for it to determine its legal legitimacy and status and the circumstances under which it could aspire to take on the regulation of commercial space transport and accept the fact that former air traffic controllers, pilots and other technical staff should not be relied upon to determine ICAO's legitimacy on this issue. It also addresses in some detail the changes that may be necessary to the existing structure of the regulation of civil aviation, both in terms of adapting the Chicago Convention principles and those of the Annexes to the Convention, when drafting a new multilateral treaty on the regulation of commercial space transportation. As part of this book, the author has drawn on and adapted some of his earlier writings as relevant to the subject of commercial space transport.

Unlike what others have provided, this book offers no magic formula. It only contains some thoughts for future consideration.

Montreal, October 2014   Ruwantissa Abeyratne

Contents

Chapter 1
Issues Involved

1.1 Air Space and Outer Space

In 1944, when the war was coming to an end, President Roosevelt invited the nations of the world to a conference in Chicago with a view to establishing a global regime for the burgeoning commercial aviation sector that would flourish after the war. The President said in his invitation: "I do not believe that the world today can afford to wait several years for its air communications. There is no reason why it should. Increasingly, the airplanes will be in existence...". At the present time, this message could be applied to commercial space transport with just a change of words to adapt to the increasingly developing space tourism sector.

It would not be true to say that ICAO has hitherto not addressed the issue of space transport. The ICAO Assembly, at its 16th Session held in Buenos Aires from 3 to 26 September 1968, adopted Resolution A 16-11 (Participation by ICAO in Programmes for the Exploration and Use of Outer Space). Recognizing that the events of the past years were of great interest to ICAO, since many of the activities affect matters falling within ICAO's competence under the provisions of the Chicago Convention, and that the United Nations had recognized the competence of certain specialized agencies that could perform various useful functions and such interest had to be welcomed and encouraged, the Assembly resolved that ICAO be responsible for stating the position of international civil aviation on all related outer space matters and for stating international civil aviation's particular requirements in respect of application of space technology.

Commercial space flight is here, and it is a foregone conclusion that it will take off in earnest within the next few years. The issue is whether we are prepared with the necessary legislative and infrastructural base to launch these flights in sustained progression. Unlike the Chicago Convention of 1944 which was adopted at the Chicago Conference pursuant to the initiative of President Roosevelt, there is no multilateral legal instrument that provides comprehensively for commercial space transportation. When commercial air transport was in its incipient stages the world

© The Author(s) 2015
R. Abeyratne, *Regulation of Commercial Space Transport*,
SpringerBriefs in Law, DOI 10.1007/978-3-319-12925-9_1

community took just over a month to develop, draft, and adopt the Chicago Convention that comprehensively provided for technical and commercial regulation. This treaty still serves air transport well.

We continue to use and explore outer space, take pictures, calculate trajectories of planets and determine who owns the moon and what the purpose of outer space exploration is. An added dimension is the use of aerospace in terrestrial transportation where an aerospace plane will take off as an aircraft, go into orbit, enter the atmosphere using the Earth's orbit into its destination, cutting the travel time significantly. It is said that by using this method, air travel time can be reduced drastically. For instance, a journey by air between Los Angeles and Sydney, which would now take 14–16 h by conventional air travel, could take 2 h or less.

It is well known that the newest and most expensive mode of transportation is commercial space travel which offers high-end suborbital flights to space tourists. Some consider it an extension of air travel, mostly on the basis that a spacecraft that operates suborbital flights would have to traverse airspace to go beyond it. This has encouraged some academics to suggest that this area of travel can easily be accommodated within the existing air transport regime, by incorporating the various safety principles that would be adopted for commercial space travel within existing treaty provisions with some adaptation and modification. Although there may be some commonality in both air travel and outer space travel, on the basis that a vehicle operating suborbital flights would go through airspace, it would be both unwise and impracticable to move a space travel regime lock stock and barrel into the existing air transport regime.

Given that a spacecraft traverses airspace before it goes into outer space, one would have to have a clear, internationally accepted definition of outer space. No multilateral treaty currently applicable to space transportation or the exploration and exploitation of outer space resources has this definition. This is both disconcerting and unsettling in an age where complex and advanced space exploration has been taking place well over 5 decades, with a man stepping on the moon in 1969. In this context, neither is airspace defined, although commercial air travel has been regulated for the past 67 years.

Given that a spacecraft traverses airspace before it goes into outer space, one would have to have a clear, internationally accepted definition of outer space. No multilateral treaty currently applicable to space transportation or the exploration and exploitation of outer space resources has this definition.[1] A State, according to the 1933 definition in the Montevideo Convention, has to be composed *inter alia* of a "defined geographic area", which is controlled by its populace. Therefore, it goes

---

[1] Another growing area of outer space activity is mining asteroids. See Ruwantissa Abeyratne, Mining Asteroids: Security Aspects, *Journal of Transportation Security*, published on line on 14 March 2013.

without saying that a State has sovereignty over its defined land area. Judge Huber noted in *the Island of Palmas* Case[2] that:

> Sovereignty in the relations between States signifies independence. Independence in relation to a portion of the globe is the right to exercise therein, to the exclusion of any other State, the function of a State. The development of the national organization of States during the last few centuries and, as a corollary, the development of international law, have established this principle of exclusive competence of the State in regard to its own territory in such a way as to make it the point of departure in settling most questions that concern international relations. Sovereignty in relation to a portion of the surface of the globe is the legal condition necessary for the inclusion of such portion in the territory of any particular State.[3]

As for territorial waters adjacent thereto (which is considered as included in the territory of a State) this has also been recognized by multilateral treaty in Article 3 of the Convention on the United Nations Law of the Sea (UNCLOS) which stipulates that "Every State has the right to establish the breadth of its territorial sea up to a limit not exceeding 12 nautical miles, measured from baselines determined in accordance with UNCLOS". Therefore, Article 2 of the Chicago Convention and its definition of "territory" is consistent with the aforementioned established principles of international law.

As per principles of public international law, air space does not constitute the territory of a State and should not be confused with the word "territory". The reason is that the dimensions of airspace have not been defined either by treaty or by customary recognition (Australia has a domestic law which recognizes that airspace goes up to 100 km over its land and adjacent sea territory. Other instances of definition of airspace are hitherto unknown). Therefore States cannot *ipso facto* claim sovereignty over the airspace above their territory unless such is recognized by treaty and this is what the Chicago Convention does in Article 1.

The Permanent Court of International Justice, when requested for a definition of "air space" in the 1933 *Eastern Greenland's Case,*[4] was of the view that the natural meaning of the term was its geographical meaning. The most fundamental assumption that one could reach from this conclusion is that air space is essentially geo-physical, meaning that it is space where air is found. Simplistically put, "air space" has been considered as going upwards into space from the territorial boundaries of a State and downwards to the centre of the Earth, in the shape of an inverted cone.

Recognition by States of their sovereignty over their airspace (as stated in Article 1 of the Chicago Convention) inevitably presupposes that this rule has already been entrenched in the annals of air law in an earlier instrument. The Convention Relating to the Regulation of Aerial Navigation signed by 26 States on 13 October

---

[2] 2 RIAA (1928) 829.

[3] *Id,* 832.

[4] PCIJ Series A/B, No. 53, at p. 53ff.

1919 established that *the High Contracting Parties recognize that every Power has complete and exclusive sovereignty over the airspace above its territory.*

As for outer space, at the time of writing, the aerospace community was considering such issues as sub-orbital flights and space tourism, both of which could further blur the boundaries between air space and outer space, while raising other issues of topical interest. As already mentioned, so far, there has not been a universally accepted definition distinguishing air space and outer space. Some years ago, when the legalities of an aerospace plane, which is a hypersonic single stage to orbit reusable vehicle that horizontally takes off and lands on a conventional runway were considered, it was thought that the transit through near space which is involved is incidental to the main transit which takes place within the airspace. Generally, the aerospace plane, which will be constructed with the use of aeronautical and space technologies and would be capable, and, indeed, required to fly both in airspace and outer space, would bring to bear the need to consider the applicability of and appropriateness of laws relating to the space plane's activities. It will be subject to the sovereignty of the State whose airspace it is in. This is an incontrovertible fact which need not be stated since any object within the airspace of a territorial State would indeed be subject to that State's sovereignty.

The United Nations Committee on the Peaceful Uses of Outer Space (UNCOPUOS), which is the UN forum where technical and legal aspects of space activities with global impact are considered, has discussed the issue of the definition and delimitation of outer space from 1962 and no definite conclusion has been reached so far in this regard. In this connection, it is of interest to note that the Legal Subcommittee of UNCOPUOS, through its Working Group on Matters Relating to the Definition and Delimitation of Outer Space, has been considering possible legal issues with regard to aerospace objects. A questionnaire thereon was circulated to all U.N. Member States. A compilation of the replies received and an analytical summary of such replies, as well as a historical summary on the consideration of the question on the definition and delimitation of outer space, may be found on the OOSA website.[5]

As debated for decades in the framework of UNCOPUOS, it may be questioned whether the vertical limit of airspace would be critical to determine the scope of applicability of air law as opposed to international space law conventions (spatialist approach), or whether the type of activities at issue would determine which law should apply (functionalist approach) to sub orbital flights. The latter school of thought submits that flights which would be passing merely in transit through (sub) orbital space in the course of an earth-to-earth transportation would be in air space and therefore remain subject to principles of air law.

A sub-orbital flight is a flight up to a very high altitude which does not involve sending the vehicle into orbit. 'Sub-orbital trajectory', which a sub orbital flight would follow, is defined in the legislation of the United States as "The intentional

---

[5] www.oosa.unvienna.org/index.html.

flight path of a launch vehicle, re-entry vehicle, or any portion thereof, whose vacuum instantaneous impact point does not leave the surface of the Earth."

In 2004, SpaceShipOne was the first private vehicle to complete two sub-orbital flights within 2 weeks carrying weight equivalent to three human adults up to about 62.5 miles (100 km) to win the Ansari X Prize. It was carried during 1 h by an aeroplane up to nearly 50,000 feet (9.5 miles) from where it was released into a glide and then propelled vertically for 80 s by a rocket motor to an altitude of more than 62 miles at apogee, reaching a speed over Mach 3. Then falling back to return to earth, it re-entered the atmosphere and glided during 15–20 min before landing back on the runway of departure.

SpaceShipOne, strictly speaking, does not operate as an aeroplane or even as an aircraft during the ballistic portion of the flight while it is not supported by the reactions of the air, even though some degree of aerodynamic control exists throughout the trajectory from launch altitude until the craft enters the upper reaches of the atmosphere where the air density is no longer sufficient for aero-dynamic flight. After apogee, during re-entry into the atmosphere the vehicle transitions to unpowered aerodynamic (gliding) flight for the return to earth. Consequently, depending upon some design and operational aspects, it could be considered operating as an aircraft in flight during this latter portion of the journey.

Therefore, such vehicles could fulfil the principal elements in the definition of aircraft and be used as such during a portion of their flights, but they offer some characteristics of a rocket as well. It is likely that other vehicles engaged in the future in such sub-orbital flights would similarly be of an hybrid nature, taking into account that developments to come may lead to a range of designs, some of which could be more clearly classified as aircraft. Should sub-orbital vehicles be con-sidered (primarily) as aircraft, when engaged in international air navigation, con-sequences would follow under the Chicago Convention, mainly in terms of registration, airworthiness certification, pilot licensing and operational requirements (unless they are otherwise classified as State aircraft under Article 3 of the Convention).

Plans have been announced by Virgin Galactic for the development of a fleet of five sub-orbital vehicles to carry paying passengers, six per vehicle; it planned that the first of these will be ready for commercial operations in 2008 at the earliest. There are indications that at least one other company is planning to offer rival sub-orbital flights.

Manned and unmanned sub-orbital flights have been undertaken to test space-craft and launch vehicles intended for later orbital flight, but some vehicles have been designed exclusively to reach space sub-orbitally: manned vehicles such as the X-15 and SpaceShipOne, and unmanned ones such as ICBMs and sounding rockets.Sub-orbital tourist flights will initially focus on attaining the altitude required to qualify as reaching space. The flight path will probably be either vertical or very steep, with the spacecraft landing back at its take-off site.

The spacecraft will probably shut off its engines well before reaching maximum altitude, and then coast up to its highest point. During a few minutes, from the point when the engines are shut off to the point where the craft begins to slow its descent for landing, the passengers will experience.

1.2 Suborbital Flights

A suborbital flight is known to be the next generation of commercial passenger travel. At the present time flight testing of commercial reusable launch vehicles (RLVs) is underway, making the availability of frequent suborbital flight closer than ever. As earlier mentioned sub orbital flights are considered missions that fly out of the atmosphere but do not reach speeds needed to sustain continuous orbiting of the earth. They allow passengers to look down at the brilliant curvature of the earth as they would from orbit.

One must not confuse a sub orbital flight with a space flight which is a flight *into* or *through* space. The craft which undertakes a spaceflight is called a spacecraft. It is often thought that orbital spaceflights are spaceflights and sub-orbital spaceflights are less than actual spaceflights. This is not entirely accurate as both orbital and sub-orbital spaceflights are true spaceflights.

The term *orbit* can be used in two ways: it can mean a trajectory in general, or it can mean a closed trajectory. The terms *sub-orbital* and *orbital spaceflights* refer to the latter: an orbital spaceflight is one which completes an orbit fully around the central body.

From the above discussion the conclusions that could be drawn are that for a flight from Earth to be a spaceflight, the spacecraft has to ascend from Earth and at the very least go past the edge of space. The edge of space is, for the purpose of space flight, often accepted to lie at a height of 100 km (62 miles) above mean sea level. Any flight that goes higher than that is by definition a spaceflight. Although space begins where the Earth's atmosphere ends, the atmosphere fades out gradually so the precise boundary is difficult to ascertain. Therefore one could argue that there is a need to accept the fact that vehicles which would effect earth-to-earth connections through sub-orbital space could incorporate the constitutive elements of aircraft and fly as such at least during descending phase while gliding. However, rocket-propelled vehicles could be considered as not falling under the classification of aircraft.

From a spatialist viewpoint, there is no clear indication in international law on the delimitation between airspace and outer space which would permit a conclusion on the applicability of either air law or space law to sub-orbital flights. On the other hand, it might be argued from a functionalist viewpoint that air law would prevail since airspace would be the main centre of activities of sub-orbital vehicles in the course of an earth-to-earth transportation, any crossing of outer space being brief and only incidental to the flight. UNCOPUOS, and more particularly its Legal Subcommittee, is considering the question of possible legal issues with regard to aerospace objects but no final conclusion has been reached yet.

1.3 Some Divergent Views

While there is a relatively well articulated legal regime governing activities of States and private entities,[6] including a regulatory arm in the form of UNCOPUOS which is a United Nations Committee answerable to the General Assembly with no law making or adjudicatory powers,[7] it has been said that the involvement of ICAO in space tourism should also be seriously considered.[8] *Jakhu* and *Battacharya* have anchored their argument in support of ICAO involvement in regulating outer space traffic on the fact that space traffic will have to be harmoniously blended with air traffic as space and air traffic management issues will be a major factor in the allocation of "air slots" to both aircraft and spacecraft. This is somewhat far-fetched, given the very high number of frequencies that may involve air travel in the future as against commercial space travel. They envision ICAO's role to be critical in the near future as a regulator and arbiter resolving space traffic issues.[9] *Jasent-uliyana*'s thinking on the regulation of outer space activities, published in 1995, introduces another dimension where he suggests that UNCOPUOS could draw on the work of ICAO:

> ...COPUOS could, by following the example of some of the specialized agencies of the United Nations, like the World Health Organization (WHO), the International Maritime Organization (IMO) and particularly the International Civil Aviation Organization (ICAO) seek to formulate international standards and recommended practices.[10]

Wassenbergh attempted in 1997 to answer the question as to whether there would be a need for a new global space organization to monitor the public law aspects of space activities or whether ICAO could act as such.[11] Although he did not answer the question directly, he offered the conclusion that regulation by a space organization would ensure the safety of the activities, orderly international competition and optimum protection of the Earth and space environment.[12]

Given its nature of work and its mandate, it would be unwise to devolve upon UNCOPUOS the responsibility of regulating space tourism. As to whether a separate Organization is created, or indeed whether regulation of space tourism is brought within ICAO will be decided as a matter of practicality. The only obstacle in the latter case is that ICAO does not have a mandate to involve itself in anything other than civil

---

[6] See Manfred Lachs, *The Development and General Trends of International Law in Our Time*, 169 Recueil Des Cours (1980) Foreword at p. xii.

[7] See Ralph G. Steinhardt, Outer Space, *United Nations Legal Order*, Volume 2 (Oscar Schachter and Christopher C. Joyner ed) American Society of International Law: 1995, 753 at p. 757.

[8] Jakhu and Battacharya 2002, pp. 112 et seq.

[9] *Ibid.*

[10] N. Jasentuliyana, A Survey of Space Law as Developed by the United Nations, published in *Perspectives on International Law*, (Nandasiri Jasentuliyana ed.) Kluwer Law International: The Hague: 1995, Chapter 16, 349 at p. 380.

[11] Wassenbergh 1997, pp. 529–535.

[12] *Id.* 535.

aviation. The aerospace community itself is confused on this issue, as could be seen in the fact that a recent study concluded that an international Space Traffic Management Organization (STM) must be established primarily for the civil and commercial use of outer space,[13] only to mention later on in the same Study that "undoubtedly, the ideal solution to accommodate space traffic management and other space safety requirements would be to amend the Chicago Convention thereby expressly extending ICAO's jurisdiction over space".[14] Be that as it may, it is incontrovertible that some ICAO involvement will be necessary, in view of the inevitable overlap between the air transport and space transport segments of a journey, particularly in the field of air traffic management. As *Jasentuliyana* suggests, the existing outer space activity regime, in terms of regulation could remain within UNCOPUOS with analogous principles attenuated from the Annexes to the Chicago Convention on subjects such as licensing, documentation to be carried on board, and rules of outer space travel. These principles should address the need for some accountability and responsibility on the part of States with regard to Standards and Recommended Practices. The closest compromise has come from *Jasentuliyana* who suggests that the existing outer space activity regime, in terms of regulation could remain within UNCOPUOS with analogous principles attenuated from aviation on subjects such as licensing, documentation to be carried on board, and rules of outer space travel. Unfortunately, it is not plausible that this Committee of the United Nations could shoulder such a large responsibility as it does not have either the competence or the resources.

Although there is a component of travel involving airspace in the overall process of space travel, they are mutually exclusive both in the consideration as to whether a person travelling as a space tourist becomes a passenger carried by air during the segment traversed in airspace and in the consideration as to whether a spacecraft can be treated as an aircraft for that segment. It is one thing to draw an analogy and take existing treaty and Annex provisions pertaining to air transport as examples but a totally incongruous thought to import all that involves outer space travel into a regime that has the sole mandate to regulate air transport. As for UNCOPUOS, this too would not work, as it is a politically driven committee that does not have the competence to handle an entirely separate and large subject as commercial air travel.

1.4 Security Implications

It is incontrovertible that, prior to any work of a policy nature commencing on the subject of commercializing space travel, a sustainable and sustained regime which governs this species of transportation has to be established. Apart from definitions of outer space and delimitation thereof, which have already been discussed in this

---

[13] *The Need for an Integrated Regulatory Regime for Aviation and Space: ICAO for Space? The Need for an Integrated Regulatory Regime for Aviation and Space—ICAO for Space?* (Jakhu, Sgobba and Dempsey ed.) Springer:2011 at p. XVII.

[14] *Id.* 138.

article, the definition of an aerospace space object,[15] the definition of a space tourist[16] the definition of a suborbital flight[17] requirements for safety and licensing and other technical issues have to be established. These are definable, predictable, structured and scientific subjects which are left to the technocrats to write into technical Annexes to a separate treaty on the subject. No multilateral treaty would carry within it such specifications to a level of minute detail. The most important, as it involves the human factor, is the security angle. What security laws would apply to misconduct and offences committed in outer space where no State has jurisdiction? Space law is grounded on the principle that outer space is the common heritage of mankind and that no State or individual can therefore claim *rights in rem* to any portion of outer space. Air law, on the other hand, is firmly entrenched in the principle of sovereignty of States, so that a State may lay claims to rights over the airspace above its territory. This essentially means that while the implementation of air law is heavily influenced by municipal law, space law is solely grounded on legal principles binding on the community of nations. Principles of public international law therefore play an exclusive part in the application of space law principles.[18]

The above notwithstanding, a State may legislate on the conduct of its national outside that State's Jurisdiction, making it punishable if that person committed an offence outside his country of nationality. In other words, customary international law would be recognized as applicable only in so far as it is incorporated into national law. In the 1906 case of *Mortensen* v. *Peters*[19] a Danish captain was convicted by a Scottish court for contravening a fishing by law (the *Herring Fishery Scotland Act 1889*) for fishing beyond the 3 mile limit recognized by international law, although the captain was fishing within the limit prescribed by law. The court held that local statute would take precedence over customary international law and that jurisdiction prescribed by international law would be destitute of effect if it were to clash with a local law that applies national jurisdiction. The *Ratio* of the

---

[15] See Claudio Zanghi Aerospace Object, *Outlook on Space Law Over the Next 30 Years:* Essays Published for the 30th Anniversary of the Outer Space Treaty (Lafferranderie and Crowther ed.), Kluwer Law International: The Hague, at pp. 115–123.

[16] See Ruwantissa Abeyratne, *Space Security Law*, Springer: 2011, 41–49.

[17] Sub orbital flights are called such because even though they enter outer space, they do not execute a full orbit around the Earth. See Michael Chatzipanagiotis, *The Legal Status of Space Tourists in the Framework of Commercial Suborbital Fligh*ts, Studies in Air and Space Law, Volume 29, Carl HeymannsVerlag: 2011 at p. 2.

[18] The *Outer Space Treaty* of 1967 in Article II provides that outer space, including the Moon and other celestial bodies, is not subject to national appropriation by claim or sovereignty by means of use or occupation, or by any other means. Furthermore the Treaty, in Article III requires that States parties to the Treaty shall carry on activities in the exploration and use of outer space, including the Moon and other celestial bodies, according to the principles of international law, including the Charter of the United Nations, in the interest of maintaining international peace and security and promoting international cooperation and understanding.

[19] (1906) 8 F.(J.) 93.

case was that the clear words of a statute bind the court even if the provisions are contrary to law. In a later case, decided in 1936, Lord Atkin stated:

> International law has no validity except in so far as its principles are accepted and adopted by our own domestic law...The courts acknowledge the existence of a body of rules which nations accept among themselves. On any judicial issue they seek to ascertain what the relevant rule is, and having found it they will treat it as incorporated into the domestic law, so far as it is not inconsistent with rules enacted by statutes or finally declared by their tribunals.[20]

This principle, when applied to conduct in outer space, clearly resonates the fact that individual States can legislate on conduct of their nationals in outer space irrespective of what principles of international air law contained in multilateral treaties might say.

Such domestic legislation might stipulate, to start with, that the State concerned shall have criminal jurisdiction over crimes committed on board or in relation to a suborbital vehicle or other spacecraft or space object within that State's territory, including national air space, even if the criminal conduct commenced within the territory of that State and continued and ended within another territory. The legislation might go on to say that the State shall have jurisdiction in the event of a crime committed on board or in relation to or with a suborbital vehicle, spacecraft or other space object, when such vehicle or object flies in international airspace or in outer space if that vehicle or object is registered in the State or if the offender is a national of that State or if the offender is a national of the State and the offence committed is also punishable under the jurisdiction of the State of registry of the vehicle in which the offence was committed.

The legislation, which can be tagged onto similar existing legislation applicable to domestic jurisdiction may go on to specify the authority and powers of the commander and items carried on board the suborbital craft.

Along with legislation, a critical factor in space security is security intelligence. Air transport could prove to be a good starting point and analogy in this regard, where significant lapses of security intelligence could prove to be disastrous.[21] However, in the punishment of offenders, Space law has more in common with maritime law than with air law, in that both deal with space that is not subject to national jurisdiction. Freedom of outer space, which lays the foundation for conduct of persons in outer space, is enshrined in Article 1 of the Outer Space Treaty of 1967, which stipulates that the exploration and use of outer space, including the moon and other celestial bodies, shall be carried out for the benefit and in the interests of all countries, irrespective of their degree of economic or scientific development, and shall be the province of all mankind. The provision also

---

[20] *Chung Chi Cheung* v. *R*, [1939] AC 160; 9 AD, p. 264. See also *Commercial and Estates Co. of Egypt* v. *Board of Trade* [1925] 1 KB 271, 295; 2 AD, at p. 423.

[21] See Ruwantissa Abeyratne, Cyber Terrorism and Aviation—National and International responses, *Journal of Transportation Security*, Springer: Published on line 31 May 2011. See also by the Same author, *Aviation Security Law*, Springer: 2012 at pp. 24–25.

requires outer space to be free for exploration and use by all States without discrimination of any kind, on a basis of equality and in accordance with international law. Finally, the provision grants free access to all States in relation to all areas of celestial bodies.

Freedom of the high seas is contained in Article 2 of the Convention on the High Seas of 1958 which provides that since the high seas is open to all nations, no State may validly purport to subject any part of them to its sovereignty. Freedom of the high seas are: the freedom of navigation, the freedom of fishing; the freedom to lay submarine cables and pipelines; and the freedom to fly over the high seas. These freedoms are recognized by the principles of public international law.

On a purely superficial comparison of the freedoms of outer space and the high seas, one can notice a general similarity in that both areas are open to mankind equally. However, the purposes for which the areas can be used are intrinsically different. For example, space law is all encompassing on the subject of the conduct of humans in outer space and celestial bodies. The Outer Space Treaty makes the sweeping statement that outer space and celestial bodies shall be open for exploration by mankind, which includes, *inter alia*, the freedom to conduct research, experiments and other forms of exploration. The Convention on High Seas, on the other hand is inclusive and therefore restrictive in forming areas of specific activity. UNCLOS has somewhat remedied this *lacuna* by adding, in Article 87(1), the freedom of scientific research *inter alia* to the already existing four freedoms.

Criminal conduct is an area where the principle of international law applicable to the High Seas lend themselves as a useful analogy to space law. Of course, the offence of piracy cannot be committed by astronauts who are sent to outer space in spacecraft belonging to a State. The offence has to be committed for private ends by persons in a private ship or craft. The offence of piracy in the high seas would nonetheless apply as an analogy to a similar offence committed by private individuals in outer space who do not represent a State as official crew members. This would cover the improbably but nonetheless possible events of the future such as a mutiny on board a commercial spacecraft carrying passengers (which is an analogy derived from shipping law). Piracy in outer space may also occur in instances where personnel of a space craft could act on the orders of a recognized government which is in gross breach of international law and which show a criminal disregard for human life.

The offence of piracy at sea and its consequences were succinctly defined by Judge Moore in his dissenting judgment in the 1927 *Lotus* Case. In the case of what is known as piracy by law of nations, there has been conceded a universal jurisdiction, under which the person charged with the offence may be tried and punished by any nation into whose jurisdiction he may come. I say "piracy by law of nations", because the municipal laws of many States denominate and punish as "piracy" numerous acts which do not constitute piracy by law of nations, and which therefore are not of universal cognizance, so as to be punishable by all nations. Piracy by law of nations, in its jurisdictional aspects, is *sui generis*. Though statutes may provide for its punishment, it is an offence against the law of nations; and as the scene of the pirate's operations is the high seas, which it is not the right or duty of any nation to police, he is denied the protection of the flag which he may carry, and is treated as an outlaw, as the enemy of all mankind—*hostis humani generis*— whom any nation may in the interest of all capture and punish.

Article 15 of the Convention on the High Seas of 1958 defines the offence of piracy as the following:

Any illegal acts of violence, detention or any act of depredation, committed for private ends by the crew or the passengers of a private ship or a private aircraft, and directed:

On the high seas, against another ship or aircraft, or against persons or property on board such ship or aircraft;

1. Against a ship, aircraft, persons or property in a place outside the jurisdiction of any State.
2. Any act of voluntary participation in the operation of a ship or of an aircraft with knowledge of facts making it a pirate ship or aircraft;
3. Any act of inciting or of intentionally facilitating an act described in sub-paragraph (1) or sub-paragraph (2) of this Article.

Mutatis mutandis, this provision would serve well as an analogy and persuasive authority in the event a similar offence committed in outer space or celestial body is examined by a competent court of any jurisdiction.

The Convention restricts the application of the offence of piracy to acts on the high seas or "any place outside the territorial jurisdiction of any state" which means essentially *in contextu* an island or "*terra nullius*" (no man's land). The latter is a fitting analogy for outer space or celestial body which is outside the jurisdiction of any State.

Article 19 of the Convention on the High Seas provides for remedial action and grants the right to any State on the high seas, or in any other place outside the jurisdiction of that State, to seize a pirate ship or aircraft, or a ship taken by piracy and under the control of pirates, and arrest the persons and seize the property on board. The courts of the State which carried out the seizure may decide upon the penalties to be imposed, and may also determine the action to be taken with regard to the ships, aircraft or property, subject to the rights of third parties acting in good faith.

The significance of the provisions of the Convention with regard to the offence of piracy, when related to similar offences which may be committed in outer space, lies in the universal application of the various limits of the offence itself. For example the offence of piracy includes acts of violence, detention or any act of depredation. This effectively brings to bear its relevance to instances such as the takeover, detention or illegal deviation of a spacecraft and therefore is quite useful as paradigm legislation for space law.

With regard to principles of tortious liability, both space law and shipping law share a common thread of international responsibility of States. Space law, by virtue of the principle that a State of registry retains jurisdiction in its space object, imputes territoriality to a State in relation to its space craft, space station or other space object. Therefore, any act committed by one State or individual representing a State which would adversely affect such space object would be comparable with an instance where the act of one State affects another in its territorial waters.

It may not be prudent to hash air transport and space transport together by amending an existing Convention, however attractive that might be as a quick fix. Both are very different fields of transport and should be covered by separate multilateral instruments, inasmuch as maritime transport is covered by UNCLOS. One may be relevant and helpful to the other in attenuating structure and analogy and in turn applying to the space transport regime, but that is as far as it should go. There should be a separate treaty on commercial space transport which should have technical Annexes, covering the broad areas of safety, security and environmental protection. The security Annexes must cover threat assessment, intelligence and prevention.

Commercial space transportation is a relatively new area that would bring people together and widen their vistas. As any other enterprise, this should flourish in an environment that protects human rights and the dignity of the person while ensuring safety and security. A treaty should therefore bring to bear laws and practices which are the glue that keeps commercial space travel secure and safe, fuelled by political will which will ignite its progress and development. The thrust of political will essentially lies in a security culture that must be visible in every State. A security culture would make States aware of their rights and duties, and, more importantly, enable States to assert them. Those who belong to a security culture also know which conduct would compromise security and they are quick to educate and caution those who, out of ignorance, forgetfulness, or personal weakness, partake in insecure conduct. They inculcate a culture of good governance and encourage States to enact legislation to this effect. Governance is the implementation of a set of principles towards the responsibility of moving towards a strategic direction. This exercise is carried out by enterprises, which are complex entities. Good governance is the interaction of the stakeholders taken to the ultimate destination of the implementation and evaluation of their interaction. It has been said that the commencement of good governance is analogous to the design of a violin. If the design is excellent, the violin would turn out to be excellent and produce excellent music. If the violin were to be used in an orchestra, the concert would probably be excellent. However, irrespective of the design and the product, the success of the performance would largely depend on the musician.

References

Jakhu R, Battacharya R (2002) Legal aspects of space tourism. In: Proceedings of the forty-fourth colloquium on the law of outer space, p 112

Wassenbergh H (1997) An international framework for private space activities. Ann Air Space Law XXII(Part 1):529–535

Chapter 2
Safety and Efficiency Issues

2.1 Safety

2.1.1 Adaptation from the Chicago Convention

Once ICAO straightens out its status and is empowered by its member States to take on the regulation of commercial space transport ICAO may wish to look at adapting the safety and efficiency principles of the Chicago Convention to space transport. The above discussions bring to bear the inexorable fact that, to start with, it is not prudent to lump air transport and space transport together by amending an existing Convention, however attractive that might be as a quick fix. Both are very different fields of transport and should be covered by separate multilateral instruments. At the least, any involvement of ICAO should be separate from its responsibilities pertaining to civil aviation under the Chicago Convention. If ICAO's reputation as a specialized agency is taken into consideration in the regulation of commercial space transport, a separate multilateral treaty could be adopted under the auspices of ICAO. Of course, as already mentioned, the title of ICAO would have to be changed to reflect reality. Separate Annexes to the new Convention pertaining to safety, environmental control and security could be developed, within reasonable time taking the already existing air Annexes as a lead on. The Organization should then be strengthened with experts in space transport and adequate resources. If there is an overlap between the two modes of transportation they would only occur in the descent phase of an aerospace vehicle at which stage existing ICAO regulations and guidelines pertaining to aviation should apply.

The separation between civil aircraft and military aircraft as contained in Article 3 of the Convention could be applied analogically to spacecraft. Article 3 bis provides that Contracting States recognize that every State must refrain from resorting to the use of weapons against civil aircraft in flight and that, in case of interception, the lives of persons on board and the safety of aircraft must not be endangered. This provision, which does not modify in any way the rights and obligations of States set

© The Author(s) 2015
R. Abeyratne, *Regulation of Commercial Space Transport*,
SpringerBriefs in Law, DOI 10.1007/978-3-319-12925-9_2

forth in the Charter of the United Nations, could be adapted to apply to the use of weapons against commercial spacecraft. Furthermore, Article 3 *bis*, which stipulates that Contracting States recognize that every State, in the exercise of its sovereignty, is entitled to require the landing at some designated airport of a civil aircraft flying above its territory without authority or if there are reasonable grounds to conclude that it is being used for any purpose inconsistent with the aims of the Convention, can be applied in principle and as adapted to spacecraft and spaceports. States may also, according to the provision, give such aircraft/spacecraft any other instructions to put an end to such violations. For this purpose, the Contracting States may resort to any appropriate means consistent with relevant rules of international law, including the relevant provisions of this Convention, specifically para (a) of Article 3 bis explained above. Each Contracting State could be required to publish its regulations in force regarding the interception of civil aircraft/spacecraft.

Article 3 *bis* which states that Contracting State shall establish all necessary provisions in its national laws or regulations to make such compliance mandatory for any civil aircraft registered in that State or operated by an operator who has his principal place of business or permanent residence in that State, can also be adapted to apply to commercial spacecraft. Each Contracting State could be required to make any violation of such applicable laws or regulations punishable by severe penalties and shall submit the case to its competent authorities in accordance with its laws or regulations.

Each Contracting State could also be required to take appropriate measures to prohibit the deliberate use of any spacecraft registered in that State or operated by an operator who has his principal place of business or permanent residence in that State for any purpose inconsistent with the aims of the Convention.

According to Article 17 of the Chicago Convention aircraft have the nationality of the State in which they are registered. This provision could be adapted to apply to commercial spacecraft. Similarly a spacecraft will not be validly registered in more than one State, but its registration may be changed from one State to another. As per Article 19 of the Convention, which can also be adapted, the registration or transfer of registration of spacecraft in any Contracting State shall be required to be made in accordance with its laws and regulations. As per Article 20, every spacecraft engaged in commercial space transportation could bear its appropriate nationality and registration marks. In accordance with the principle in Article 21, each State could be required to undertake to supply to any other Contracting State or ICAO, on demand, information concerning the registration and ownership of any particular spacecraft registered in that State. In addition, each Contracting State can furnish reports to ICAO, under such regulations as the latter may prescribe, giving such pertinent data as can be made available concerning the ownership and control of spacecraft registered in that State and habitually engaged in commercial space transportation. The data thus obtained by ICAO shall be made available by it on request to the other States.

Article 29 on documents on board can be adapted to reflect that every spacecraft engaged in commercial space transport will be required to carry in conformity with the conditions prescribed in the new Convention on Commercial Space Transport: its certificate of registration; its certificate of space-worthiness; the appropriate licenses for each member of the crew; its journey log book; if it is equipped with radio and other apparatus, the appropriate license; a list of their names and places of embarkation and destination; and a manifest and detailed declarations of the cargo. Every spacecraft engaged in commercial space transport could be required to be provided with a certificate of airworthiness issued or rendered valid by the State in which it is registered. Article 32 of the Chicago Convention can be adapted to require that the commander of every commercial spacecraft and the other members of the operating crew of every spacecraft engaged in commercial air transport shall be provided with certificates of competency and licenses issued or rendered valid by the State in which the spacecraft is registered. Certificates of space-worthiness and certificates of competency and licenses issued or rendered valid by the Contracting State in which the spacecraft is registered, shall be recognized as valid by the other Contracting States, provided that the requirements under which such certificates or licences were issued or rendered valid are equal to or above the minimum standards which may be established from time to time pursuant to the Convention.

Article 34 of the Convention can be validly adapted to provide that in respect of every spacecraft engaged in commercial space transport a journey log book in which shall be entered particulars of the spacecraft, its crew and of each journey, in such form as may be prescribed from time to time pursuant to the Convention. Another important provision of the Chicago Convention to be adapted to commercial space travel is Article 37 which can provide that each Contracting State undertakes to collaborate in securing the highest practicable degree of uniformity in regulations, standards, procedures, and organization in relation to spacecraft personnel, space paths and auxiliary services in all matters in which such uniformity will facilitate and improve commercial space transportation. The exception provided in Article 38 can be adapted to read that any State which finds it impracticable to comply in all respects with any such international standard or procedure, or to bring its own regulations or practices into full accord with any international standard or procedure after amendment of the latter, or which deems it necessary to adopt regulations or practices differing in any particular respect from those established by an international standard, shall give immediate notification to the International Civil Aviation Organization of the differences between its own practice and that established by the international standard. In the case of amendments to international standards, any State which does not make the appropriate amendments to its own regulations or practices shall give notice to the Council within 60 days of the adoption of the amendment to the international standard, or indicate the action which it proposes to take. In any such case, the Council shall make immediate notification to all other States of the difference which exists between one or more features of an international standard and the corresponding national practice of that State.

Article 39 is also valid which says that which there exists an international standard of airworthiness or performance, and which failed in any respect to satisfy that standard at the time of its certification, shall have endorsed on or attached to its airworthiness certificate, a complete enumeration of the details in respect of which it so failed. It goes on to say that any person holding a license who does not satisfy in full the conditions laid down in the international standard relating to the class of license or certificate which he holds, shall have endorsed on or attached to his license, a complete enumeration of the particulars of any aircraft or part thereof with respect to in which he does not satisfy such conditions. Furthermore another adaption of the Convention could be the provision that can say that no spacecraft or personnel having certificates or licenses so endorsed shall participate in commercial air transportation, except with the permission of the State or States whose territory is entered. The registration or use of any such spacecraft, or of any certificated spacecraft part, in any State other than that in which it was originally certificated shall be at the discretion of the State into which the spacecraft or part is imported.

Article 83 *bis*, on the transfer of registration of aircraft in the Convention can read in the new Convention as when an spacecraft registered in a Contracting State is operated pursuant to an agreement for the lease, charter or interchange of the spacecraft or any similar arrangement by an operator who has his principal place of business or, if he has no such place of business, his permanent residence in another Contracting State, the State of registry may, by agreement with such other State, transfer to it all or part of its functions and duties as State of registry in respect of that spacecraft. The State of registry could then be relieved of responsibility in respect of the functions and duties transferred. The transfer shall not have effect in respect of other Contracting States before either the agreement between States in which it is embodied has been registered with the Council and made public pursuant to Article 83 or the existence and scope of the agreement have been directly communicated to the authorities of the other Contracting State or States concerned by a State party to the agreement.

2.1.2 Lessons Learned from Air Transport

When looking at the regulation of safety of commercial space transport, one would do well to look at problems of safety caused by air transport due to economic wrangling. In order to devise successful safety management systems, it is necessary to identify the exigencies of air transport that might pose threats to safety. Operations carried out by foreign registered aircraft reflects the first area of concern regarding safety. Within the basic principle enunciated in Article 17 of the Chicago Convention, which provides that aircraft have the nationality of the State in which they are registered, air operators have, over the past two decades, increasingly employed foreign registered aircraft for various reasons. Aircraft being leased or otherwise interchanged and operated outside the State of Registry, sometimes for long periods of time, is common. There are also instances where a foreign registered aircraft might be leased or sub-leased or chartered from one country to another. While such arrangements are legitimate from an economic and regulatory perspective, they can present problems from a safety viewpoint because of the bifurcation of the State of Registry and State of the Operator. For example, this could result in a situation where operators can be subject to different interpretations of SARPs of ICAO as implemented by different States.

A major safety concern is the problem of "flags of convenience" associated with foreign registered aircraft. "flags of convenience" is a term derived from the maritime industry which denotes a situation in which commercial vessels owned by nationals of a State, but registered in another State, are allowed to operate freely between and among other States. When an aircraft rarely, if ever, returns to the State of Registry, its airworthiness oversight becomes an issue in the absence of safety oversight arrangements between the State of Registry and the State of the Operator. There are broadly two groups of foreign registered aircraft that can be deemed to operate under a flag of convenience: those done for fiscal purposes and those done to take advantage of a system with minimal economic or technical oversight or no oversight at all. The first group may not pose a serious problem if arrangements are made between concerned States to ensure proper oversight, for example through bilateral agreements under Article 83 *bis*, which permits States to transfer all or a part of certain safety oversight responsibilities under the Convention. Even for this group, the reality remains far from satisfactory in that relatively few bilateral agreements implementing Article 83 *bis* have been notified to ICAO (by 1 May 2006, 148 States had ratified the provision), and numerous aircraft of all types all over the world are still subject to split oversight responsibility. It is the second group that creates a major safety problem which needs to be addressed.

Another area that might impinge on the safety of flight concerns operations involving foreign flight crew. Split oversight problems could also occur in respect of foreign-licensed flight crew. For example, dry leases (i.e. the lease of an aircraft without crew) raise the problem of validation of foreign crew licences by the State of Registry. The issue becomes complicated when the rules and requirements for crew licences in the State of Registry are at variance with the corresponding rules in the State that initially issued the licences. Differences between the laws and regulations of the State of Registry and those of the State of the Operator may also exist in the case of wet leases (i.e. a lease of aircraft with crew). While the lessor usually remains the official operator in such cases, the lessee may already operate aircraft of a similar type under its Air Operators Certificate. It may happen then that the wet-leased aircraft are operated under the lessee's Air Operator's Certificate and, consequently, the State of the lessee becomes the State of the Operator. In such circumstances, proper surveillance of the operating crew may become difficult. The situation could become more complicated if the operation involves a mixed crew (e.g. the cabin crew from the lessee carrier and the cockpit crew from a foreign lessor carrier).

"Off-shore" operations involving flight operations away from the designating State, (State of Registry or State of the Operator) may also have safety implications. In a situation where the designated airlines of a bilateral agreement are granted the so-called 7th freedom rights (i.e. *to carry traffic from the second State to/from third State(s) without the need for the service to connect the home State*), such airlines may set up an operational base in a second country for services to/from third countries. Where cabotage or right of establishment is permitted, air carriers may operate in the territory of the granting State. Such a situation could raise the question as to how the required safety oversight should be handled between the State of the Operator and the State in which the operation is based.

Growing commercial trends in air transport suggest that operations involving multiple parties and the use of another's brand, such as code sharing and franchising, could adversely affect safety unless checked by the States concerned. Code sharing has been the most prevalent element in transnational airline alliance arrangements and can take a variety of forms. Although it is usually treated as a commercial arrangement, because of the complexity of some code sharing arrangements (e.g. a flight using the codes of several carriers from different countries), the safety/security authorities may find it difficult to determine their level of involvement *vis-à-vis* other authorities. In such circumstances, the questions of responsibility and accountability for safety/security can lead to uncertainty. Also, since such arrangements allow an operator to use the name or assume the public face of another carrier (e.g. in the case of franchising), the need to safeguard reputation in terms of service/safety quality have led to some regulatory action on safety/security. For example, some States require foreign airlines with which their national airlines have code sharing arrangements to meet a similar level of safety. This could also raise a question as to whether all States whose airlines are involved in a code sharing operation should be involved in such safety oversight, and to what extent each should be involved. Another concern arising from code sharing relates to the security implications caused by the potential transfer of a security threat, which may exist against one airline and be spread to its partner or partners in a code sharing arrangement, and any subsequent additional security measures imposed by the appropriate authorities. Since technical and operational regulations may vary considerably from one partner airline/State to the other, this raises the question as to how the accountability and responsibility for safety/security should be handled amongst the partner airlines and States.

Cross-border airline merger/acquisition, where allowed, could lead to such companies having operations or places of business in different States, or operating mainly outside the State in which their registered offices and/or owners are located. This situation could raise questions regarding the attribution of regulatory oversight responsibility amongst the States concerned (e.g. in the case of the merged airline having two principal places of business), or on the application of whose standards, where they differ between the countries concerned.

Outsourcing of activity affecting aircraft operation is now a regular practice in the air transport industry and can take various forms. For example, airlines may outsource their ground handling; send their aircraft to be repaired and/maintained in foreign countries, and contract out certain flight operations and/or crew administration to another airline or company. In each of these cases, multinational industries have emerged to provide such services. Some States also encounter situations where an Air Operator's Certificate applicant had only a corporate skeleton with most of the proposed operational activities to be performed/provided by foreign companies (including the aircraft and flight crews). This situation could present challenges for the licensing and safety oversight authorities from both the State issuing the Air Operator's Certificate and the State of the outsourced activity on how to ensure that such practice or entity properly meet the safety and security requirements.

While some of the above situations already make it difficult individually for identifying or attributing the responsibility for safety/security compliance and oversight, it could become even more problematic when dealing with a complex situation that combines many or all of the above features. As reflected in the above discussion, there is an increasing number of situations in which one is dealing with a cascade of States, each having a share of responsibility in an air transport operation. The challenge for States is how to ensure that, regardless of the form of regulatory or commercial arrangement, there should always be a clear point of contact for the safety and security oversight responsibility in a clearly identified State or its delegated authority for any given aircraft operation.

2.2 Efficiency

2.2.1 Fuel Efficiency

Article 44(d) of the Chicago Convention states that one of ICAO's aims is to meet the needs of the people of the world for safe, regular, efficient and economical air transport. Understandably, if ICAO were to be involved in the regulation of commercial space transport these four factors would apply in that context too. In the aviation context, the focus is on fuel efficiency as well as efficiency in air navigation and it could well be that the Secretary General of ICAO, in his letter to Commander Hadfield meant the same for commercial space transport. However, the ambivalence reflected in the word "efficiency" in the letter of the Secretary General, which could even mean efficiency in ground support is confusing. He could have explained in his letter exactly what he meant by "efficiency". One commentator says of efficiency in ground support:

> A ground system is the network of facilities that support a spacecraft during the various phases of mission development. The functions provided by a ground system range from testing of spacecraft functionality during its integration, through prelaunch performance confirmation, to the following post launch operations tasks: planning and scheduling, spacecraft command and control, navigation, and data collection, analysis, assessment, and distribution. To perform these functions, ground systems typically use a wide array of items including generalized and specialized spacecraft test equipment, communications hardware and software, and computers with sophisticated specialized software for control and data gathering, analysis, and assessment[1]

---

[1] See Whitworth et al. 1998, pp. 247–256 at 247.

Of course, in the area of fuel efficiency in commercial space transport much headway has been made, where engineers of the National Aeronautics and Space Administration (NASA) of the US have finished testing a new ion-propulsion system for earth-orbiting and interplanetary spacecraft. The system is more powerful and fuel-efficient than its predecessors, enabling it to travel farther than ever before. Ion propulsion works by electrically charging, or ionizing, a gas using power from solar panels and emitting the ionized gas to propel the spacecraft in the opposite direction. The concept was first developed over 50 years ago, and the first spacecraft to use the technology was *Deep Space 1* in 1998.[2]

In the aviation context too headway has been made. Over the past decades, market forces have always ensured that fuel burn (and associated CO_2 emissions) has been kept to a minimum for efficiency reasons. As a result of permanent fleet modernization, with new aircraft achieving unmatched efficiency performance, fuel burn has been reduced by about 70 % over the last 40 years.

Improvements in aircraft fuel efficiency are inextricably linked to how engine, aircraft and systems manufacturers design their products. The concepts, the design criteria, the design optimization and the technology transition processes are all tightly interconnected, and the interactions usually increase as a product is developed. Generation after generation, aircraft have shown impressive weight reductions, aerodynamics improvement and engine performance increase, thus reducing drastically the amount of fuel burn (and of CO_2 emitted) to perform the same or further improved operational mission.

Simultaneously, product innovations are permanently introduced through design, simulation, modelling, testing and validation tools. The optimization process and the challenging tradeoffs involve iterative loops at the technology, design and product levels. As a consequence, it takes approximately a decade to design and develop an aircraft. In order to make the appropriate decisions, when investing in future technologies, aircraft engine and airframe manufacturers need a stable regulatory framework, based on dependable scientific knowledge, and consistent funding to sustain the current and future extensive research programmes.

Some commitments have been made by the manufacturing industry to keep that improvement trend: in Europe, the goals set by the Advisory Council for Aeronautics Research in Europe (ACARE) are targeting an additional 50 % improvement in fuel burn and associated CO_2 emissions in 2020, compared to 2000 performance. This should be done while reducing the perceived noise levels by 50 % and the emissions of NO_x by 80 % over the same period. Comparable objectives are set in the US, through the different programmes running with the NASA for instance.

Associated research programmes, development clusters to foster synergies through appropriate partnerships, have been set up, thus enabling to better take in consideration the challenges associated to the interdependencies between environmental improvement and other parameters (performance, economics…) and within

---

[2] See http://www.technologyreview.com/news/414620/a-more-efficient-spacecraft-engine/.

the environmental criteria themselves between noise, local air quality and climate change-related issues. Some choices for future technology engine and aircraft configuration will be based on the decisions the society will make relative to these challenges.

In addition to the traditional embedded technological improvement pattern that aircraft engine and airframe manufacturers are continuously supporting, some further opportunities to further reduce the emissions from aviation may arise. They are related to the design, development, validation and production of sustainable alternative fuels. This new avenue must be further explored to identify the environmental benefit it can generate, on top of the expected technological improvements.

Fuel efficiency is a major consideration for aircraft operators as fuel currently represents around 20 % of total operating costs for modern aircraft. One of the options being considered for aviation, where emissions are released in high altitude affecting the environment and impacting local air quality, is the use of alternative fuels. The most serious concern lies in the impact of emissions caused by the currently used fossil fuels to climate change, which has prompted the Kyoto Protocol (1997)[3] to the United Nations Framework on Climate Change (UNFCCC) to require developed countries in particular to decrease their collective green house gas emissions by 5 % by the period 2005–2012, with 1990 as a bench mark. The demand for air travel continues to increase, making the concern of aircraft engine emissions caused by fossil fuel burn even more compelling.[4] Preliminary results from ICAO's Committee on Aviation Environmental Protection (CAEP) indicate that the demand for air travel is expected to continue to grow through at least 2036 and on a per-flight basis; efficiency is expected to continue to improve throughout that period. The anticipated gain in efficiency from technological and operational measures is not expected to completely offset the predicted growth in demand driven emissions, leaving a potential "mitigation gap" to achieving sustainability. Commercializing sustainable alternative fuels for aircraft can be an essential strategy for closing this gap. While there are no significant quantities of such fuels available for commercial aviation today, it is anticipated that these fuels will become an essential component of the future aircraft fuel supply.

---

[3] The Protocol was deemed to enter into force 90 days after "not less than 55 Parties to the [Climate Change] Convention, incorporating Parties included in Annex 1 which accounted in total for at least 55 % of the total carbon dioxide emissions for 1990 of the Parties included in Annex 1" have ratified (Article 24 of the Protocol).

[4] The world uses 3,917 Mega tonnes (Mt) of liquid fuel annually 3, including approximately 0.02 Mt of bio fuel, very little of which is consumed by international aviation. Most fuel use is for direct combustion, emitting carbon dioxide (CO_2) in direct proportion to fuel burn. Preliminary estimates from ICAO's Committee on Aviation Environmental Protection (CAEP) indicate that global aviation fuel burn is expected to grow from approximately 200 Mt in 2006 to between 450 and 550 Mt in 2036. Not accounting for the impact of alternative fuels, but considering the effects of improved efficiency and aircraft technologies, CO_2 is predicted to grow from 632 Mt in 2006 to the range of 1,422–1,738 Mt in 2036.

The world uses 3,917 mega tonnes (Mt) of liquid fuel annually, of which 0.02 Mt is biofuel. Very little of this biofuel is used by international aviation. By 2036, international aviation could need a substantial contribution from sustainable alternative fuels for aircraft in order to reduce its overall GHG footprint. One of the many significant benefits in the use of sustainable alternative fuels in aircraft would be the drastic lowering of emissions containing sulphur oxides (SO_x) and particulate matter (PM),[5] as a result of which local air quality would be improved.

Removing sulphur from conventional jet fuel to produce a low sulphur jet fuel will significantly reduce PM and SO_x emissions from aircraft. Currently, jet fuel has a specification maximum of 3,000 parts per million (ppm) for sulphur; however, jet fuel in the market has a lower sulphur content. Worldwide surveys conducted during 2007 found that annual weighted average jet fuel sulphur content ranged from 321 to 800 ppm.[6] Hydrodesulphurization, which could be applied to remove the fuel sulphur, is a common process in petroleum refineries, and low sulphur diesel fuel is already widely used internationally. Low sulphur jet fuel has sulphur content less than 15 ppm.

As regards PM, According to a recent analysis of the impacts of aviation emissions on human health,[7] primary PM emissions are responsible for 13 % of total PM impacts. Secondary PM is much more significant with sulphur-related PM emissions responsible for 33 % and NO_x-related PM emissions responsible for 54 %. With low sulphur jet fuel, SO_x emissions would be significantly reduced, which in turn would result in a significant reduction in secondary PM. Overall, volatile primary PM emissions are reduced due to reduced fuel sulphur content. Hydrodesulphurization causes other fuel modifications that also reduce non-volatile PM emissions.

The current understanding of experts with regard to PM pollution is insufficient to fully evaluate the magnitude of health and environmental effects of exposure. However, indications are that the size of PM is a significant factor. Coarse particles can be inhaled but tend to remain in the nasal passage. Smaller particles are more likely to enter the respiratory system. Health studies have shown a significant association between exposure to fine and ultra fine particles and premature death from heart or lung disease. Fine and ultra fine particles also have been linked to effects such as cardiovascular symptoms, including cardiac arrhythmias and heart attacks, and respiratory symptoms such as asthma attacks and bronchitis. These effects can result in increased hospital admissions, emergency room visits, absences from school or work, and restricted activity days. Individuals that may be particularly sensitive to fine particle exposure include people with heart or lung disease,

---

[5] Particulate matter (PM) from fuel combustion is a mixture of microscopic solids, liquid droplets, and particles with solid and liquid components suspended in air. Solid particles, such as soot or black carbon, are referred to as non-volatile particles. Volatile PM is comprised of inorganic acids (and their corresponding salts, such as nitrates and sulphates), and organic chemicals from incomplete fuel combustion.

[6] See Taylor 2009.

[7] See Brunelle-Yeung 2009.

older adults, and children.[8] Standard setting organizations specify requirements that jet fuel must meet for physical properties, chemical content, contaminant limits, and overall performance requirements. To limit PM and SO_x fuel combustion emissions, fuel standards have maximum limits on the sulphur content of fuels.

2.2.2   Efficiency in Navigation

The Global Navigation Satellite System (GNSS) will be relevant to commercial space transportation and ICAO's considerations on efficiency in space travel will revolve round the fundamental nature and use of the system. The ICAO Assembly, at its 32nd Session in 1998, adopted Resolution A32-19 (Charter on the Rights and Obligations of States Relating to GNSS Services) containing fundamental principles of a Charter of Rights and Obligations of States in relation to GNSS Services.[9] In the Resolution, States, whilst recognizing that the primary use of GNSS services is to maintain safety in international civil aviation, reaffirm the principle that every State and aircraft of all States shall have access, on a non-discriminatory basis, and under uniform conditions to the use of GNSS services. The Resolution also grants every State authority and control over aircraft operations over their territory, and, *inter alia* imposes obligations on provider States to ensure continuity, availability, integrity, accuracy and reliability of such services.

ICAO is spearheading efforts to share best operational practices in close-in procedures such as CDA (Continuous Descent Arrival) that reduces holding patterns and thus emissions. Similarly, ICAO fosters PBN (Performance Based Navigation) which results in more direct routing thereby saving fuel and reducing aviation emissions. Efficiency of air transport is inextricably linked to environmental protection by the supply and demand curve pertaining to air transport services. Public expectations and demand coupled with the traffic growth are increasing pressures on the air navigation system to, inter alia, put more aircraft in the skies, reduce delay, be environmentally sustainable, and minimize operational costs, and to do so safely. In order to assist States and regions, ICAO developed the global air traffic management (ATM) operational concept which reflects a global vision of an integrated, harmonized and interoperable PBANS up to and beyond 2025. This vision is supported by the Global Air Navigation Plan (GANP), which recognizes that near-term performance gains could be achieved through more effective planning and implementation of existing technologies, procedures and capabilities It is contingent on ICAO and other stakeholders to ensure the consistent delivery of the GANP and its set of common and harmonized initiatives to States

---

[8] U.S. Environmental Protection Agency, Review of the National Ambient Air Quality Standards for Particulate Matter; Policy Assessment of Scientific and Technical Information, http://www.epa. gov/ttn/naaqs/standards/pm/data/pmstaffpaper_20051221.pdf, December 2005.

[9] Res A32-19, Charter on the Rights and Obligations of States Relating to GNSS Services. *Assembly Resolutions in Force* (as of 5 October 2001) Doc. 9790, at V-3.

and regional planners. ICAO must continue to strive for global consensus and provide guidance on solutions that will, *inter alia*, address the introduction of new types of aircraft, and the expanded use of aircraft capabilities, thereby continuing the transition towards the global ATM.

One particular issue relevant to the aviation industry and ICAO is that significant near-term efficiency and environmental gains may be achieved by the deployment of new technologies such as: replacement of expensive infrastructure through the use of new technologies while maintaining or exceeding the existing level of performance; development of new procedures that take advantage of aircraft capabilities; shift from a static paper-based information system to a real-time and paperless information environment; development and implementation of more accurate weather forecasts.

It is at this point that the problem arises, and legal discourse begins. The mere fact that the Charter is now an ICAO Assembly Resolution has prompted the comment:

> Adopted in the form of an Assembly Resolution, the Charter cannot be accorded any legal force and therefore must be regarded as legally not binding. Some commentators, having expressed serious doubts as to the usefulness of the instrument, seem to be somewhat displeased with the nomenclature employed which would be indicative of a legal instrument of fundamental importance[10].

This is seemingly consistent with another view on the ICAO Assembly which states that:

> [ICAO] Assembly recommendations... are more than hortatory. They are designed to set global norms in a field where there is widespread acknowledgment of the need for ordered conduct. They are adopted by a plenary body, with the shared expectation that States will follow them to the extent that they are able to. They clearly are not binding, but they have a sufficient channelling effect to place them well above the low point on a continuum of normative instruments ranging from non law to true law.[11]

However, unlike the former view, this statement attributes more coercive force to ICAO Assembly Resolutions, thus making the Charter on States' rights and obligations relating to GNSS an instrument which sets out norms and demands States to follow them if possible and is therefore in between "non law" and "true law". What this means is that the Charter would not be totally destitute of effect in establishing certain obligations for States to perform. Therefore, the Charter becomes a reckonable force in international relations if not at international law, particularly since ICAO resolutions are highly persuasive and carry much political leverage. Above all, such an instrument could, while reaffirming existing legal principles, pave the way for an international convention that is binding on States' Parties.

The significance of the legal status regarding the current principles on the conduct of States in using space based applications in air traffic management lies

---

[10] See Andrade 2001, p. 89.

[11] See Kirgis 1995, Chapter Eighteen, pp. 825 at 840.

in the compelling need to inquire as to whether rigid fragmentation of law and policy is really necessary, particularly in such an important area as aviation safety. Does one dismiss policy in this critical area purely on the inflexible notion that it is not enforceable? On the other hand, do States need to abandon rigid demarcations in instances such as these and agree to global adherence?

The 32nd Session of the ICAO Assembly also adopted Resolution A32-20 (Development and Elaboration of an Appropriate Long Term Legal Framework to Govern the Implementation of GNSS). This resolution, which recognizes that GNSS is an important element of the ICAO CNS/ATM System, is aimed at providing a framework governing safety-critical services for aircraft navigation with worldwide coverage. It also recognized, *inter alia*, the need for an appropriate long term legal framework to govern the implementation of GNSS and endorsed the Council decision taken earlier to authorize the ICAO Secretary General to establish a Study Group on Legal Aspects of CNS/ATM systems. The Assembly instructed the Council and the Secretary General to consider the elaboration of an appropriate long term legal framework to govern the operation of GNSS including consideration of an international convention.

Resolution A32-20 was a signal development in global recognition of the need to inquire into a liability regime regarding damage caused by the GPS process through signals transmitted to air navigation facilities. The Resolution resulted in an ICAO Secretariat Study Group being established to elaborate proposals for a liability framework. The Study Group reported to the 33rd Session of the Assembly (held in September–October 2001) in a somewhat divided way, some members recommending that the applicable regime under domestic law was adequate and appropriate to cope with the global navigation satellite system and others being of the view that a global international law instrument, such as a Convention, might be required to address issues of liability in the long term. A compromise between these two views, representing a model contractual framework, was also suggested to the Assembly as an alternative. The Assembly further remanded the matter to the Study Group to finalize the concept of a contractual framework as a first step, with a view to later considering the development of an international convention as a long term measure.

At the ICAO Eleventh Air Navigation Conference, held in Montreal from 22 September to 3 October 2003, the European Organization for the Safety of Air Navigation (EUROCONTROL), on behalf of its members and those of the European Civil Aviation Conference (ECAC), presented such a contractual framework to the Conference.[12] The framework contained main elements developed while taking into consideration regional requirements. The Conference was advised that, in the face of increased advancement and development of GNSS, there was a compelling need for an adequate legal and institutional framework to cover

---

[12] GNSS Legal Framework—Contractual Framework for the Implementation, Provision, Operation and Use of the Global Navigation Satellite System for Air Navigation Purposes, AN-Conf/11-WP/153, 18/9/03.

liability issues, *inter alia*, concerning GNSS. The main reason for this need, as was identified in support of the proposal, was that, as most States would not have direct involvement in the operation of the GNSS System or sub systems, air navigation services provided within their sovereign airspace will be compelled to rely on facilities which were beyond their control. It was also contended by EUROCON-TROL and others presenting the contractual framework that States needed to be satisfied, *inter alia*, that there was an appropriate and adequate performance level of the GNSS signals and services offered within their airspace, ensuring their integrity, reliability, accuracy and continuity and that such services should come within a clearly defined liability framework.

Taking into consideration the above, and with a view to filling a gap which hitherto existed, a framework agreement was proposed containing common provisions for implementation that could be used in private law contracts between parties. These provisions were calculated to address safety, certification, liability and jurisdictional issues. The framework agreement's main purpose was to ensure the existence of general principles that would ensure the ultimate responsibility of a State under Article 28 of the Chicago Convention for GNSS implementation in their territory. Certain mandatory common elements, such as: the need to comply with applicable Standards and Recommended Practices (SARPs) of the applicable Annexes to the Chicago Convention and ICAO Resolutions; fault based liability; compulsory risk coverage; recourse to arbitration; waiver of the right to invoke sovereign immunity; and ICAO's central role as global coordinator, formed the roots of the framework.

The African States who presented an alternate proposal to the Conference of an international convention to cover liability issues of GNSS,[13] reiterated the European concern that there was a large constituent of user States who by themselves could not provide GNSS services (as required by Article 28 of the Chicago Convention) but were nonetheless responsible for the provision of those services under international treaty. As such, it was the view of the African States that an international convention was needed in order to establish and adopt principles regarding provider and user responsibility and rights in the event of damage. Such a Convention, according to the African States, would also serve to provide for an international institutionalized safety and security oversight process. The African States further suggested that the Convention should be based on ICAO's central responsibility to develop principles and techniques and foster planning and development of international air transportation, with safety as a paramount concern. The suggested Convention would involve mandatory submission to arbitration by parties concerned with an accident or damage and oblige providers to assure continuity, availability, accuracy and transparency while being liable for damage caused by GNSS services provided by them.

---

[13] Legal Aspects of GNSS, AN-Conf/11-WP/143, 18/9/03.

The United States, on the other hand, recommended[14] to the Eleventh ICAO Air Navigation Conference that there was no need for either a new contractual framework or international convention. The United States was of the view that ICAO has flexible legal tools (such as SARPs) together with a sustained and long term institutional framework and experience that would enable the Organization to continue to serve without major problems, as it had, through such major aviation landmarks as the advent of the jet engine, radar and many other scientific advancements without legal or institutional problems. As such, the United States invited the Conference to agree that work on GNSS has progressed steadily over the past several years and no deficiencies had been found to impede technical implementation of CNS/ATM and that such work should not be impeded or delayed by work on legal and institutional issues.

A fourth school of thought, represented by the LTEP, was inclined to tie in the responsibility of States under Article 28 of the Chicago Convention to adherence to SARPs. The LTEP suggested that States providing signals in space or under whose jurisdiction such signals are provided should certify the signals in space by attesting that such provision is in conformity with SARPs. The LTEP went on to recommend that the States having jurisdiction under the Chicago Convention should ensure that avionics, ground facilities and training and licences requirements comply with ICAO SARPs.[15] These recommendations were based on the LTEP's considered view that, irrespective of whether or not a State handed over the provision of air navigation services to a privatized entity, ultimate responsibility continued to devolve upon the State for the provision of such services under Article 28 of the Convention.[16] The LTEP favoured an international convention on liability as an ultimate and long term measure, taking into consideration that a GNSS induced incident may, in certain circumstances, involve multiple and complex actions in several jurisdictions, thus requiring an internationally harmonized legal regime containing a simple, clear and speedy procedure.[17]

One of the material issues that the LTEP had to address involved seeking some commonality in the different legal regimes applying in the various jurisdictions. For example, the law applicable in the United Kingdom was based on a fault liability system whereas the law in France was based on the principles of administrative law.[18] On the other hand, the law in Canada regarding GNSS was governed by a hybrid system of common law and civil law whereas issues of liability in the United States were dealt with under the common law system in accordance with the

---

[14] Legal and Institutional Issues and the Status of CNS/ATM, AN-Conf/11-WP/160, 18/9/03.

[15] Recommendation 3 of LTEP. See Report of the Secretariat Study Group on Legal Aspects of CNS/ATM Systems, First Meeting, SSG-CNS/1-Report 9, April 1999 at p. 4.

[16] *Id.* p. 3.

[17] *Id.* p. 5.

[18] See Report of the Secretariat Study Group on Legal Aspects of CNS/ATM Systems, Second Meeting SSG-CNS/2—Report, 9 November 1999, p. 2.

principles of tort law.[19] The laws of Italy and Australia were essentially founded on delictual (or tortuous) principles of liability.

The LTEP presented its findings to the ICAO Council at the 170th Session of the Council in November 2003. In its Report,[20] the LTEP advised the Council that the Panel had finalized the draft text of a *Contractual Framework Relating to the Provision of GNSS Services* containing obligations of the GNSS signal provider as well as those of the air traffic service provider which makes use of the signal for the provision of its services. Essentially, the draft contractual framework foresees a series of contracts between the various stakeholders while stipulating that the signal provider is obligated to provide the signals with regularity, continuity, integrity, accuracy and uninterrupted availability.[21]

With regard to the perceived need for an international convention as a long term measure, the LTEP advised the Council that, should a Convention be adopted, its framework should not constitute a non-tariff barrier for market access and more importantly, should not impede future technical innovations. Furthermore, it was contended that any proposed framework for a Convention must not restrict the freedom of parties of entering or terminating services. The overall submission of the LTEP to the Council was that, in view of the paucity of experience with regard to the implementation of CNS/ATM and GNSS in particular, it was premature to embark on drafting an international convention. The LTEP suggested that the Panel continue its work with a view to finalizing its remaining work.

References

Andrade AAL (2001) The global navigation satellite system. Ashgate, Aldershot, p 89

Brunelle-Yeung E (2009) The impacts of aviation emissions on human health through changes in air quality and UV irradiance. Thesis, Master of Science in Aeronautics and Astronautics, Massachusetts Institute of Technology, Boston, MA, May 2009

Kirgis FL Jr (1995) Aviation. In: Schachter O, Joyner CC (eds) United nations legal order, vol 2. Cambridge University Press, Cambridge

Taylor WF (2009) Survey of sulphur levels in commercial jet fuel. CRC Aviation Research Committee of the Coordinating Research Council, Alpharetta, GA, February 2009

Whitworth GG, Somers AJ, Stratton WC (1998) Efficient spacecraft test and operations with the NEAR ground system. Johns Hopkins APL Tech Dig 19(2):247–256

---

[19] *Id.* pp. 3–4.

[20] Progress Report on the Work of the Secretarial Study Group on the Legal Aspects of CNS/ATM Systems, C-WP/12080, 7/11/03.

[21] *Id.* p. 3.

Chapter 3
Economic and Social Issues

3.1 Economic Issues Relating to Commercial Space Transport

3.1.1 ICAO's Views on Economic Oversight

The Secretary General of ICAO, in his letter of invitation to Commander Hadfield (alluded to in the Preface) speaks of addressing economic and social issues related to commercial space transport at the UN/ICAO Symposium. In other words, how would commercial space transport be regulated economically. Regrettably, ICAO's track record in the economic aspects of air transport has been abysmal[1] and the mind boggles as to how ICAO will have the credibility to address this issue. Article 44 of the Chicago Convention[2] merely cites as an aim and objective of ICAO fostering the development of air transport. Would ICAO discuss at the symposium the same "fostering" of commercial space transport? If, as some have simplistically stated that a separate Annex to the Chicago Convention can be drawn up for commercial space transport, would that Annex be subject analogically to Article 6 to the Convention which stipulates that no scheduled international air service may be operated over or into the territory of a contracting State, except with the special permission or other authorization of that State, and in accordance with the terms of such permission or authorization?

Another aspect for consideration would be economic oversight of commercial space activities. How would ICAO view spaceports and navigation service providers? At the Conference of ICAO on the Economics of Airports and Air Navigation Services (CEANS), which was held in Montreal from 15 to 20 September 2008,[3] The ICAO Secretariat, in submitting its views on economic oversight to the

---

[1] See generally, Ruwantissa 2013.
[2] *Supra* note 1.
[3] CEANS was attended by 520 delegates from 104 States and 19 international organizations.

© The Author(s) 2015
R. Abeyratne, *Regulation of Commercial Space Transport*,
SpringerBriefs in Law, DOI 10.1007/978-3-319-12925-9_3

Conference, commenced with the fundamental postulate that the State is ultimately responsible for protecting the interests of users through economic oversight defined the term "economic oversight" as monitoring by a State of the commercial and operational practices of service providers.[4] It was suggested that economic oversight may take several different forms, from a light-handed approach (such as the reliance on competition law) to more direct regulatory interventions in the economic decisions of service providers. It is interesting that the Secretariat took a direct and clear position that States may perform their economic oversight function through economic regulation, either through legislation or rule-making, and/or the establishment of a regulatory mechanism.[5]

It was also argued that the objectives of economic oversight could include: ensuring that there is no abuse of dominant position by service providers; ensuring non-discrimination and transparency in the application of charges; providing incentives for service providers and users to reach agreements on charges; ensuring that appropriate performance management systems are developed and implemented by service providers and assuring investments in capacity to meet future demand. The priority for each objective may vary depending on the specific circumstances in each State, and there should be a balance between such public policy objectives and the efforts of the autonomous/private entities to obtain the optimal effects of commercialization or privatization.

The Conference was advised that there were already several modalities in Doc 9082,[6] para 15 of which recommends that States establish an independent mechanism for the economic regulation of airports and air navigation services. This provision suggests that such a mechanism would oversee economic, commercial and financial practices and its objectives could be drawn or adopted from, but need not be limited to certain principles.[7] The Secretariat also drew the attention of the Conference to the *Manual on Air Navigation Services Economics*[8] and the *Airports Economics Manual*[9] which suggest such modalities of economic oversight as (a) application of competition law; (b) fallback regulation, whereby regulatory interventions are limited to situations when the behaviour of the regulated entity breaches publicly-stated acceptable bounds; (c) institutional arrangements such as

---

[4] Economic Oversight, CEANS-WP/4, 16/4/08.

[5] *Id.* 1.

[6] ICAO's Policies on Charges for Airports and Air Navigation Services, Doc 9082 5th Edition, 2009.

[7] The principles alluded to in para 15 are: ensure non-discrimination in the application of charges; ensure there is no overcharging or other anti-competitive practice or abuse of dominant position; ensure transparency as well as the availability and presentation of all financial data required to determine the basis for charges; assess and encourage efficiency and efficacy in the operation of providers; establish and review standards, quality and level of services provided; monitor and encourage investments to meet future demand; and ensure user views are adequately taken into account.

[8] *Manual on Air Navigation Services Economics*, Doc 9161/3, Third Edition, 1997.

[9] *Airport Economics Manual*, Doc 9562, Second Edition: 2006.

requirements on consultation with users (often supplemented by arbitration/dispute resolution procedures), information disclosure, and a particular ownership, control and financial structure; (d) a third-party advisory commission, whereby a group of interested parties reviews pricing, investment and service levels proposals; (e) contract regulation, whereby the State grants a contract, or concession, to provide airport or air navigation services under certain conditions; (f) incentive-based or price-cap regulation; and (g) cost of service or rate of return regulation.

3.1.2 Other Views

During CEANS, one delegation suggested that regional organisations can provide the necessary resources for States that do not have their own capacity to adequately perform economic oversight functions. It recommended that there should be mechanisms for ICAO to work with such regional organizations through the development of guidance material. Another delegation put forward the view that there was a compelling need for regulatory interventions to be measured and applied in a manner proportionate to the specific circumstances. Yet another delegation underscored the need for economic regulation and urged States to implement the ICAO Assembly Resolution A 36-15[10] regarding economic regulation of international air transport.

Delegates were unanimous in the view that economic oversight of airports and air navigation services is a necessary State responsibility with the promotion of an appropriate balance amongst safety, security and facilitation, environmental and economic issues. The overall package of economic instruments should provide net economic benefits for all developing countries and preferential measures for the Least Developed Countries in particular. There was also the view of one delegation that the role of the States in economic oversight in the form of legislation or through the establishment of an appropriate regulatory mechanism to resolve the issues on the increase of the cost of aviation fuel was vital while another stressed that applying similar forms of economic oversight to airports and ANSPs ignores the differences between the two types of service providers, in particular their divergent degree of competition. Therefore, it was contended that the proposed amendment of Doc 9082 should be consistent with the underlying assumption that airports do not *per se* have a dominant market position. The suggestion was also made that any regulatory interventions should be kept at a minimum, be subject to a cost-benefit analysis, and ensure sufficient investment to meet future demand.

---

[10] A 36-15, Consolidated Statement of continuing policies in the air transport field, *Assembly Resolutions in Force*, Doc 9902, at III-I.

3.1.3 Conclusions of the Conference

There was discussion during CEANS where some delegations suggested that, in order to "give teeth" to ICAO policy, there be a recommendation in Doc 9082 to the effect that amendment to Doc 9082 should be incorporated by States in their national legislation. It is submitted that such a measure would tantamount to treading uncharted and dangerous ground. While it is one thing to assert that the only way that ICAO policies could be implemented is for States to opt incorporating such principles in their legislation, it is something quite different to recommend that States go ahead and do so.

As a necessary compromise and in order to reach a balance, the Conference broadly recognized the need for economic oversight in the increasingly commercialized and privatized environment for airports and air navigation services. It considered a number of suggestions that were made by the delegates for improving the proposed new text for Doc 9082. The following conclusions were reached by the Conference:

(a) States should bear in mind that economic oversight is the responsibility of States with the objectives, *inter alia*, to prevent the risk that a service provider could abuse its dominant position, to ensure non-discrimination and transparency in the application of charges, to encourage consultation with users, to ensure the development of appropriate performance management systems, and to ascertain that capacity meets current and future demand, in balance with the efforts of the autonomous/private entities to obtain the optimal effects of commercialization or privatization;

(b) States should select the appropriate form of economic oversight according to their specific circumstances, while keeping regulatory interventions at a minimum and as required. When deciding an appropriate form of economic oversight, the degree of competition, the costs and benefits related to alternative oversight forms, as well as the legal, institutional and governance frameworks should be taken into consideration;

(c) States should consider adoption of a regional approach to economic oversight where individual States lack the capacity to adequately perform economic oversight functions; and

(d) ICAO should amend Doc 9082 to clarify the purpose and scope of economic oversight for airports and air navigation services with reference to its different forms and the selection of the most appropriate form of oversight.[11]

---

[11] Draft Report on Agenda Item 1.1., Economic Oversight, CEANS-WP/73, 16/9/08, Draft Report on Agenda Item 1.1.

3.2 Social Issues

In the inevitable event of commercial space transport becoming prolific in the years to come, it is incontrovertible that chances of encountering extra terrestrial life would be more of a possibility than before. Marcia S. Smith[12] envisions that if life were to be found in outer space the ensuing conduct of human kind would be more an ethical issue. Any inquiry into life in outer space has to take into account both vegetation as well as intelligent life. In both instances the main consideration would be how we could protect such life forms and use them for the benefit of human kind. The central theme of space exploration would incontrovertibly be international cooperation and abstinence from the use of force, which collectively form the cornerstone of space exploration from a legal standpoint.

Ms. Smith asks the pertinent questions: "Do we send more probes to further investigate and do we have a responsibility to protect that life and allow it to develop naturally? If robotic probes definitively find life, should we erect a "do not disturb sign" and refrain from sending further probes?" While these questions would have to be asked and answered at one point or another, the more immediate issue would be what we would do on Earth to cope with the new exigency.

One of the corollaries to finding life in outer space would be the issue of how we would use such a discovery in the context of the prevailing environment of international relations. In this context international politics within the umbrella of the United Nations and the United Nations Charter may become extremely relevant. It is not unrealistic to envision that the discovery of life in outer space could spark a discourse on interests and a renewed initiative to revisit international treaties to ensure the peaceful uses of outer space while at the same time ensuring some degree of control on the use of life so discovered.

At the 79th Plenary Meeting of its 61st Session, the United Nations General Assembly adopted Resolution 61/111[13] which, *inter alia*, expresses serious concern of the General Assembly about the possibility of an arms race in outer space and urges all States, in particular those with major space capabilities, to contribute actively to the goal of preventing an arms race in outer space as an essential condition for the promotion of international cooperation in the exploration and use of outer space for peaceful purposes. Doubtless, such a threat would prove to be more real and ominous if life in outer space were to be discovered.

The General Assembly also agreed that a panel of space exploration activities, including the participation of the private sector should be convened during the 50th session of the United Nations Committee for the Peaceful Uses of Outer Space (UNCUPUOS). Perhaps the most noteworthy of the Assembly's observations as

---

[12] Smith 2006, pp. 217 at 221.

[13] A/RES/61/111, *International Cooperation in the Peaceful Uses of Outer Space*, 15 January 2007.

recorded in the Resolution is that the recommendations of UNISPACE III[14] could be integrated into the work programme of the Office of Outer Space Affairs and that UNCUPUOS could consider these recommendations for implementation. UNISPACE III is the genesis of the Vienna Declaration[15] which, *inter alia* espouses the protection of the outer space environment.

3.2.1 Is There Life in Outer Space?

Although the title of this article is purely conjectural, its contents—in the eventuality of the occurrence suggested therein becoming real—bring to bear the need to reflect on principles of human conduct and liability prescribed by existing legal norms and ethical considerations with regard to the treatment of life.

American geneticist Joshua Lederberg introduced to the world the science of exobiology (or astrobiology)—a branch of biology which deals with the search for extraterrestrial life, especially intelligent life, outside the solar system. Although remote astronomical observations of a planet or other celestial body provide information about its physical environment, the determination of the presence of life on these bodies is more difficult. Exobiological techniques are designed to detect life forms, artefacts produced by intelligent life, waste produces of metabolic reactions, remnants of former life, pre-biological molecules that may reflect early evolutionary stages or substances such as carbon or combination of Hydrogen and Oxygen forming water that are necessary for the sustenance of life as it is experienced on Earth.

In 1948 the U.S. Air Force commenced maintaining a file of reports relating to extraterrestrial phenomena called Project Blue Book. In July 1952, the U.S. government established a panel of scientists including engineers, meteorologists, physicists and an astronomer to investigate a series of radar detection coincident with visual sightings near the national airport in Washington D.C. The panel was organized by the Central Intelligence Agency, which underscores the thrust of public and government concern and interest at the time.[16]

The concern was based on U.S. military activities and intelligence and that its report was originally classified Secret. Later declassified, the report revealed that 90 % of UFO sightings could be readily identified with astronomical and

---

[14] The Space Millennium: Vienna Declaration on Space and Human Development, adopted by the Third United Nations Conference on the Exploration and Peaceful uses of Outer Space (UNISPACE III) held at Vienna from 19 to 30 July 1999. See Report of the Third United Nations Conference on the Exploration and Peaceful uses of Outer Space, United Nations Publication, Sales No. E.00.1.3. See also, A/RES/59/2 (30th Plenary Meeting, 20 October 2004) and A/RES/60/99 (62nd Plenary Meeting, 8 December 2005).

[15] *Ibid.*

[16] See *The New Encyclopaedia Britannica*, Volume 12, Micropaedia, 15 ed. Encyclopaedia Britannia Inc. Chicago, 1990 at p. 129.

meteorological phenomena (e.g. bright planets, meteors, auroras, ion clouds) or with aircraft, birds, balloons, searchlights, hot gases, and other phenomena, sometimes complicated by unusual meteorological conditions.

The publicity given to early sightings in the press undoubtedly helped stimulate further sightings not only in the U.S. but also in Western Europe, the Soviet Union, Australia, and elsewhere. A second panel established in February 1966 reached conclusions similar to those of its predecessor. This left a number of sightings admittedly unexplained, and in the mid-1960s a few scientists and engineers, notably James E. McDonald, a University of Arizona meteorologist, and J. Allen Hynek, a Northwestern University astronomer, concluded that a small percentage of the most reliable UFO reports gave definite indications of the presence of extra-terrestrial visitors.

This sensational hypothesis, promoted in newspaper and magazine articles, met with prompt resistance from other scientists. The continuing controversy led in 1968 to the sponsorship by the U.S. Air Force of a study at the University of Colorado under the direction of E.U. Condon, a noted physicist. The Condon Report, "A Scientific Study of UFO's" was reviewed by a special committee of the National Academy of Sciences and released in early 1969. A total of 37 scientists wrote chapters or parts of chapters for the report, which covered investigations of 59 UFO sightings in detail, analyzed public-opinion polls and reviewed the capabilities of radar and photography. Condon's own "Conclusions and Recommendations" firmly rejected ETH—the extraterrestrial hypothesis—and declared that no further investigation was needed.

This left a wide variety of opinions on UFO's. A large fraction of the U.S. public, and a few scientists and engineers, continued to support ETH. A middle group of scientists felt that the possibility of extraterrestrial visitation, however slight, justified continued investigation, and still another group favoured continuing investigation on the grounds that UFO reports are useful in socio-psychological studies. These varying views and attitudes were expressed at a symposium held by the American Association for the Advancement of Science, in December 1969. Several years later, in 1973, a group of U.S. scientists organized the Centre for UFO Studies in Northfield, Ill., to conduct further work.

On 24 June 1947, American pilot Kenneth Arnold described some unusual flying craft he had seen over the mountains off the west coast of the United States. In his words: "they flew like a saucer would if you skipped it across water".[17] Newspapers applied his phrase to the craft themselves, and the misleading label "flying saucer" has followed the phenomenon of the unidentified flying object (UFO) since.

The Word UFO—unidentified flying object—officially means simply something that has not been, or cannot be accounted for by any of the known laws of physics. But the seemingly rational behaviour reported in many UFO sightings, as well as

---

[17] Arnold and Palmer 1952 at p. 3.

the accounts of meetings with humanoids, has led to the speculation that UFOs are, in fact, spacecraft bringing creatures from outer space.

If this is so, the spaceships must be able to cover immense distances. People who claim to have had contact with extra-terrestrials often say they have spoken with Venusians. But Venus is highly unlikely to be inhabited. Any intelligent life forms must be coming from still further away, and, even assuming that life spans of creatures from other planets may be much longer than our own, it is clear that UFOs must be able to travel very fast indeed if they are not to take hundreds of years to travel between inhabited planets.

Reports of the movement of UFOs are remarkably consistent. Most people describe them as hovering and then taking off at very high speed, often executing manoeuvres that would be impossible in conventional aircraft. Even allowing for exaggeration by excited witnesses, the consistency of the reports suggests the UFOs use a very powerful force to produce dramatic accelerations.

None of the rocket fuels we use at present can produce either the speed or acceleration observed in the UFOs, because they store only a small amount of energy for a given mass. Right from the beginning, rocket travel has been faced with the problem of enabling the rocket to carry enough fuel for its journey—it must lift the fuel, which can be very heavy if the journey is long, as well as itself and its occupants. The solution has been the multi-stage rocket: the initial acceleration is given by a rocket that is jettisoned when its fuel is used up and a second rocket takes over.

Space flights have always stretched our rocket technology to its limits—and, as everyone knows, our rockets and spacecraft do not accelerate very briskly away from the Earth. Although they eventually reach quite high speeds, they are nowhere near fast enough to reach planets outside our Solar System within a human lifetime.

If we assume UFOs are subject to the same laws of physics as we are, then, to operate on and near the Earth with the rapid accelerations and manoeuvres at high speeds that are often reported, they must be using a different source of energy from conventional chemical fuels. Their fuel must be highly compact, with a high energy yield for a small mass: the obvious source is nuclear fuel.

As long ago as 1958—just after Yuri Gagarin had become the first man ever to orbit the Earth in space—Freeman Dyson, a theoretical physicist, embarked on a plan for a nuclear-powered spaceship. He had previously worked on the development of the atom bomb and had a comprehensive understanding of nuclear power. He assembled a group of scientists at La Jolla, Southern California, to work with him; he called his scheme 'Project Orion'.

Project Orion was a serious attempt to build a spacecraft powered by nuclear explosions, and was intended as an alternative to the multi-stage rockets that Werner von Braun was proposing for space travel.

Freeman Dyson's ultimate aim was to build a spacecraft the size of a small city that would take a group of people to a distant comet on the edge of the solar system, where they would settle. This may have been only a pipe-dream, but the design was real enough.

The spacecraft was to be powered by hydrogen bombs. Essentially, his idea was to carry a number of hydrogen bombs aboard the spacecraft; these would be moved, one by one, to a position underneath the craft where they would be exploded. The base of the spacecraft would absorb the shock and the craft would be driven along. Obviously the spacecraft and the bomb system would have to be designed so that the craft was propelled along and not simply blown apart, but—in principle, at least —this was straightforward. However, Dyson was never able to test his ideas: he was prevented by public concern about the pollution of the atmosphere by radio-active fallout.

UFOs are often reported as disappearing rapidly—going of 'like a television set' and reappearing just as quickly. This aspect of the phenomenon has puzzled scientists for a long time and has led to suggestions that UFOs use some kind of 'anti-optic device' to prevent them from being seen. There are, however, some simpler explanations that account for the majority of reports. UFOs 'disappearing' in the of night could do so by simply switching off their lights' daytime discs could appear to vanish by turning themselves sideways on to the observer—it would be very difficult to pick out the thin edge of a disc against the sky. These explanations do not, of course, account for radar-visual sightings that suddenly vanish. But if a UFO disappeared behind a patch of disturbed air, a mirage-like effect could easily screen it both from sight and from radar detectors.

There are, however, cases on file for which none of these explanations seem credible. It seems that the phenomena involved can only be explained as products of a technology much further advanced than our own.

By far the majority of UFO reports describe the strange objects as disc- or cigar-shaped and it could be that most UFOs reported as cigar-shaped are in fact discs. Whether or not this is actually the case, the number of reports of saucer-shaped UFOs is overwhelming. There has been a great deal of speculation as to why this should be so—some people have suggested the mystical significance of the circle may have something to do with it—but there is a simple explanation.

On long inter-stellar voyages, a spacecraft will pass through vast regions of empty space—far from the regions of gravitational attraction of any major objects —where there is no wind resistance, no up or down, no east or west, nothing. The most logical shape for a vessel travelling in these circumstances is circular, for a circle is symmetrical about an infinite number of axes. The fact that most UFOs are disc-shaped rather than spherical can be explained as a design feature that allows spacecraft to operate at high speeds once they have entered the atmosphere of planets: by flying with their edges into the wind, they can cut down the effect of air resistance almost to zero.

Whether or not UFOs existed in the past, there is no doubt that UFO sightings have proliferated in astonishing numbers over the past 30 years. This fact seems to be in some way linked with man's first steps towards exploring space, and this connection is undoubtedly an important clue in trying to explain the UFO.

Estimates of the total number of UFO sightings vary so widely as to be meaningless; more helpful figures are provided by the catalogues of reported sightings prepared by individual investigative organizations. In the 1960s a French team catalogued more than 600 encounter cases in France alone, each vouched for by responsible investigators. In the early 1970s UFO investigators made lists of all reported landing cases for particular countries: 923 were recorded in the United States, 200 in Spain.

Are UFOs real in the sense that, say, spacecraft are real? The surest proof would be actually to get hold of one, and there are persistent rumours that certain governments, notably that of the United States, have indeed obtained a UFO, which is kept in total secrecy. However this remains mere conjecture, despite the sworn affidavits of alleged witnesses. Indeed, the whole matter of governmental involvement—or the lack of it—is a further and fascinating aspect of the UFO controversy.

In the absence of a real UFO that we can touch and examine, there is a great deal of evidence of the phenomenon in the form of a mass of photographs and a handful of movies. The majority are undoubtedly fakes. Those with good credentials are so blurred, so distant or so ambiguous that they simply add a further dimension to the problem: why, if UFOs exist, and in an age when many people carry cameras with them most of the time, have we not obtained better photographic evidence?

Perhaps the strongest evidence we have is from the effects caused by UFOs on surrounding objects, particularly machinery. In November 1967 a truck and a car approaching each other on a Hampshire road in the United States in the early hours of the morning simultaneously suffered engine failure when a large egg-shaped object crossed the road between them. The police, and subsequently the Ministry of Defence, investigated the incident, but no official explanation was ever issued. Such a case may leave investigators puzzled, but it makes one thing certain: if they can cause physical effects, UFOs must be physically real.

If they are physical objects, UFOs must originate from somewhere. When the first UFOs of the current era were seen, back in the 1940s, it was assumed they came from somewhere on Earth. The Americans suspected they were a Russian secret device, perhaps developed using the expertise of German scientists captured at the end of the Second World War.

But as more reports came in it became clear that no nation on Earth could be responsible. Nor was there sufficient evidence to support other ingenious theories—that they came from the Himalayas, long a favoured source of secret wisdom, or Antarctica, where unexplored tracts of land and climatic anomalies provide a shaky foundation for speculation. Instead, ufologists began to look beyond the Earth, encouraged by the fact that our own space explorations programme was just beginning. We were starting to take an active interest in worlds beyond, and it seemed reasonable that other civilizations might have a similar interest in us.

However, although the number of potential sources of life in the Universe is virtually infinite, the probability of any civilisation being at a stage of development appropriate for space travel is very small. They fact that no solid evidence has been found for the extraterrestrial hypothesis is discouraging. Although it is the best available explanation, it remains no more than speculation.

3.2.2 Close Encounters

Established science has always tended to view the UFO phenomenon with scepticism. In his book, *The UFO experience*,[18] Dr. J. Allen Hynek, who was astronomical consultant to Project Blue Book (the U.S. Air Force investigation into UFOs), tells the story of an event at an evening reception held in 1968 in Victoria, British Colombia, at which a number of astronomers were present. During the evening it was announced that strange lights—possibly UFOs—had been spotted outside. Dr. Hynek continues: "The news was met by casual banter and the giggling sound that often accompanies an embarrassing situation." And, he reports, not a single astronomer went outside to look.

Even project Blue Book attempted to explain away every reported sighting in terms of conventional science. It soon began to earn itself a bad name because many of its explanations were impossible to believe. In 1966 the U.S. Air Force set up a 2-year research project—to investigate, in effect, its own investigations!

Dr. J. Allen Hynek, while acting as a consultant to Project Blue Book, developed a system of classification of UFO 'types', which has become standard. He divided UFO reports according to the distance, greater or less than 500 ft. (150 m), at which the UFO was observed, and subdivided each of these two sections into three, giving six categories altogether.

The commonest sightings are of the "distant" type. *Nocturnal lights* are strange lights seen at a distance in the night sky, often with unusual features such as variations in the intensity of light or colour and sudden, remarkable changes of speed and direction of movement. *Daylight discs* are distant objects seen against the sky during the daytime. The shapes vary considerably: cigars, spheres, eggs, ovals and pinpoints as wee as discs are often reported. *Radar-visuals* are distant UFOs recorded simultaneously on radar and visually with good agreement between the two reports. Dr. Hynek excluded "sightings" made solely by radar since false traces can result from a number of natural factors such as ground scatter—the signal is reflected from high ground—temperature inversions and even thick banks of cloud or flocks of birds. Radar-visual sightings are the most important categories of UFO reports as they given independent instrumental evidence of the sighting; unfortunately, they are very rare.

Reports of UFOs seen at close range are the most interesting and often spectacular; these are the famous "close encounters". *Close encounters of the first kind* are simple observations of phenomena where there is no physical interaction between the phenomena and the environment. *Close encounters of the second kind* are similar to the first kind except that physical effects on both animate and inanimate matter are observed. Vegetation may be scorched or flattened, tree branches broken, animals frightened or car headlights, engines and radios doused. In cases of electrical failure the equipment usually begins to work normally again once the UFO has disappeared. *Close encounters of the third kind* occur when "Occupants"

---

[18] Hynek 1972 at p. 12.

are reported in or around the UFO. Dr. Nynek generally ruled out so-called "contactee" cases in which the reported claimed to have had intelligent communication with the "occupants", arguing that such reports were almost invariably made by pseudo-religious fanatics and never by "ostensibly sensible, rational and reputable persons." But even these cases occasionally have to be taken seriously by scientists.

3.2.3 Legal Principles

In a purely forensic sense, the fundamental postulate of space law, which devolves upon States the responsibility to explore and use outer space for peaceful purposes is pivotal to the conjecture of extra terrestrial life.

The first source at hand is international treaty law and the *Outer Space Treaty*[19] of 1967 is the first point of reference. The basic principle of space law is the "common interest" (or common heritage) principle which emerged as a result of the first specific Resolution on space law of the United Nations General Assembly in 1958.[20] The "common interest" principle has since been incorporated in subsequent multilateral treaties, particularly the *Outer Space Treaty* Article 1(1) which provides:

> [T]he exploration and use of outer space, including the moon and other celestial bodies, shall be carried out for the benefit and in the interest of all countries, irrespective of their degree of economic or scientific development, and shall be the province of all mankind.

Article II of the Treaty states that outer space, including the Moon and other celestial bodies, *are not subject to national appropriation* by claim or sovereignty, by means of use or occupation or by any other means. Article III follows through with the requirement that States Parties to the Treaty shall carry on activities in the exploration and use of outer space, including the Moon and other celestial bodies, *in accordance with international law,* including the Charter of the United Nations, in the interest of maintaining international peace and security and promoting international cooperation and understanding.

In 1994 the United Nations General Assembly adopted Resolution 49/34,[21] which *inter alia* covers international co-operation in the peaceful uses of outer space, and links the importance of international co-operation in developing the rule of law, including the relevant norms of space law and their important role in international co-operation for the exploration and use of outer space for peaceful

---

[19] Treaty on Principles Governing the Activities of States in the Exploration and Use of Outer Space, Including the Moon and Other Celestial Bodies, January 27 1967, 610 U.N.S.T. 205 (entered into force on 10 October 1967).

[20] *UNGA Resolution 1348 (XII)*, 13 December 1958.

[21] See U.N.G.A. Res 49/34 of 9 December 1994, Official Records of the General Assembly, 49th Session, at p. 91.

purposes. The Resolution also expresses *inter alia* the concern of the General Assembly of the United Nations with regard to the possibility of an arms race in outer space and urges COPUOS to continue its consideration of the legal aspects related to the application of the principle that the exploration and utilization of outer space should be carried out for the benefit and in the interests of all States, taking into particular account the needs of developing countries.[22] The Committee is also requested to continue to consider, as a matter or priority, ways and means of maintaining outer space for peaceful purposes and to report thereon to the General Assembly.[23]

The operative criterion for the adoption of the above aims, as identified by the Resolution, lies in the endorsement that they should be achieved through international co-operation in the development of the role of law. The Resolution therefore brings to bear the ineluctable and compelling need for the application of existing principles of international law as a means towards this end.

The General Assembly of the United Nations, on 5 February 1996, adopted Resolution 56/27[24] relating to international co-operation in the peaceful uses of outer space. This resolution broadly reaffirmed the principles of Resolution 49/34. Having considered the Report of UNCOPUOS on the work of its 38th session, the General Assembly endorsed the Committee's recommendation that the Committee should, through its Scientific and Technical Sub-Committee, *inter alia*, consider the use of nuclear power sources in outer space and questions relating to space transportation systems and their implications for future activities in space.[25] Matters relating to life sciences, space medicine and astronomy were some of the areas that were focused on for further consideration in the Resolution.[26]

The mandate which UNCOPUOS has received from the United Nations General Assembly includes the task of studying the nature of legal problems which may arise from the exploration of outer space. Other international bodies, such as the International Telecommunications Union (ITU) and the International Atomic Energy Agency (IAEA)—the latter which is interested in environmental problems and questions relating to nuclear energy—are also involved in the corollaries to the peaceful use of outer space as they affect the human being.

---

[22] *Id.* para 4(c).

[23] *Id.* para 38.

[24] U.N.G.A. A/RES/50/27, 5 February 1996, Official Records of the General Assembly, 50th Session, 96-76447.

[25] *Id.* para 17(a) (iv), 17(b) (I).

[26] *Id.* para 17(b) (iii) and (vi).

3.2.4 Peaceful Uses of Outer Space

There is no known principle or pronouncement of law which mentions "extraterrestrial intelligence". Therefore *ex facie*, no existing norm of international law prohibits social intercourse in outer space between humans and extraterrestrial beings (if such were to exist). The only prohibition which will obtain at international law would be if such intercourse leads to the "non peaceful" use of outer space by States and such conduct would adversely affect the interests of humankind. It is incontrovertible that the absence of peaceful use of outer space (when states indulge in activities in outer space) would inevitably mean warlike or aggressive use of outer space. Accordingly, such action would perforce form "use of force" by such states on other states or persons affected by these actions.

The "law" on the use of force that has developed through the United Nations is substantial. As Article 1, paras 1 and 2, of the UN Charter suggest, the primary purposes of the United Nations are: "(1) To maintain international peace and security, and to that end: to take effective collective measures for the prevention and removal of threats to the peace, and for the suppression of acts of aggression or other breaches of the peace, and to bring about by peaceful means, and in conformity with the principles of justice and international law, adjustment or settlement of international disputes or situations which might lead to a breach of the peace"; and "(2) To develop friendly relations among nations based on respect for the principle of equal rights and self-determination of peoples, and to take other appropriate measures to strengthen universal peace."

The United Nations Charter contains numerous provisions which are relevant to the use of force. Several General Assembly resolutions, adopted without dissent or with near unanimity, have restated, amplified and clarified the meaning of these Charter provisions. Possibly, the most relevant and authoritative is the 1970 Declaration on Principles of International Law Concerning Friendly Relations and Co-operation Among States in Accordance with the Charter of the United Nations (Friendly Relations Declaration).[27] Although some dispute exists in regard to the precise legal status of the *Friendly Relations Declaration*, it is generally regarded as an authoritative interpretation of broad principles of international law expressed in the Charter.[28] Another, and more controversial example, is the General Assembly's "Definition of Aggression" resolution.[29]

The UN Charter does not directly address the question of intervention by states; rather, under Article 2(7) it precludes the Organization itself from intervening "in matters which are essentially within the domestic jurisdiction of any state." Hence the General Assembly's Declaration on the Inadmissibility of Intervention in the Domestic Affairs of States and the Protection of Their Independence and

---

[27] GA Res. 2625, UN GAOR, 26th Sess., Supp. No. 28, at p. 121, UN Doc. A/8028 (1971).

[28] See Rosenstock 1971, pp. 713, 714.

[29] GA Res. 3314, UN GAOR, 29th Sess., Supp. No. 31, at p. 142, UN Doc. A/0631 (1975).

Sovereignty,[30] adopted in 1965 by a vote of 109 to none with one abstention,[31] takes on special legal significance. The Declaration, *inter alia*, provides:

> no State has the right to intervene, directly or indirectly, for any reason whatever, in the internal or external affairs of any other State. Consequently, armed intervention and all other forms of interference or attempted threats against the personality of the State or against its political, economic and cultural elements are condemned.[32]

The Declaration further provides:

> no State shall organize, assist, foment, finance, incite or tolerate subversive, terrorist or armed activities directed towards the violent overthrow of the regime of another State, or interfere in civil strife in another State.[33]

The Security Council and the General Assembly are the primary organs of the United Nations that have responsibility under the Charter to regulate the use of force. Sharp debate has arisen over the precise allocation of authority in this area between the Council and the Assembly, and a focal point of this debate has been the Assembly's "Uniting for Peace" resolution[34] whereby the Assembly claims the authority to recommend collective measures in situations where the Council is unable to deal with a breach of the peace or act of aggression in the face of a veto. A useful source which reflects the Charter's allocation of powers between the Assembly and the Council is the International Court of Justice's advisory opinion in the *Certain Expenses case*,[35] which, while not directly addressing the legality of the Uniting for Peace resolution, contains dictum relevant to this issue and generally constitutes an authoritative discussion of the Charter's allocation of powers between the Council and the Assembly. Similarly, the Court's advisory opinion in the Namibia case[36] extensively discusses the scope of decision-making authority in both organs. Even when Council and Assembly decisions are not regarded as binding, they may be considered as interpretations of the Charter entitled to some weight.

The "case law" of the United Nations on the use of force consists primarily of Security Council resolutions. In such cases as Korea, Rhodesia, South Africa, and Iraq, among others, Security Council resolutions have developed the law of the Charter.

---

[30] GA Res. 2131, UN GAOR, 20th Sess., Supp. No. 14, at p. 11, UN Doc. A/6014 (1966).

[31] The United Kingdom, see UN GAOR, 20th Sess., First Comm., at pp. 430–31, UN Doc. A/C.1/ SR 1422 (Dec. 20, 1965).

[32] Declaration on the Inadmissibility of Intervention in the Domestic Affairs of States and the Protection of Their Independence and Sovereignty, GA Res. 2131, UN GAOR, 20th Sess., Supp. No. 14, at p. 11, UN Doc. A/6014 (1966).

[33] *Id.* Article 2.

[34] GA Res/337A, UN GAOR, 5th Sess., Supp. No. 20, at p. 10, UN Doc. A/1775 (1951).

[35] Certain Expenses of the United Nations, 1962 ICJ 151 (Advisory Opinion).

[36] Advisory Opinion of the Continued Presence of South Africa in Namibia (South West Africa), 1971 ICJ 16.

Certain interpretations or legal opinions by the Secretary-General on the use of force also may carry some weight. As the chief administrative officer of the United Nations,[37] the Secretary-General is perforce obliged to carry out functions assigned him by the principal organs of the Organization, including the General Assembly and the Security Council.[38] He has no independent lawmaking authority. His responsibility is rather to implement resolutions of the deliberate organs pursuant to their terms, which, however, may at times be cast in the most general language. In such cases the Secretary-General must interpret the resolution himself. Secretary-General Dag Hammarskjold, for example, in implementing Security Council resolutions that authorized him to provide the Government of the Congo with military assistance, created considerable controversy with his interpretations of the authority conferred upon him by these resolutions. In defending these interpretations Hammarskjold later wrote:

> Is he [the Secretary-General] entitled to refuse to carry out the decision properly reached by the organs on the ground that the specific implementation would be opposed to positions some Member States might wish to take as indicated, perhaps, by an earlier minority vote? Of course, the political organs may always entrust him to discontinue the implementation of a resolution, but when they do not instruct him and the resolution remains in effect, is the Secretary-General legally and morally free to take no action, particularly in a matter considered to affect international peace and security? Should he, for example, have abandoned the operation in the Congo because almost any decision he made as to the composition of the force or its role would have been contrary to the attitudes of some Members as reflected in debates and maybe even in votes, although not in decisions?

> The answers seem clear enough in law; the responsibilities of the Secretary-General under the Charter cannot be laid aside merely because the execution of decisions by him is likely to be politically controversial. The Secretary-General remains under the obligation to carry out the policies as adopted by the organs. The essential requirement is that he does this on the basis of his exclusively international responsibility and not in the interest of any particular State or group of States.[39]

Several decisions and advisory opinions of the International Court of Justice bring to bear their relevance to the use of force. Although the doctrine of *stare decisis* does not apply to ICJ decisions, and such decisions do not create law in the common law sense, they nonetheless constitute an authoritative statement of the law and are a crucially important part of the international legal process.

A significant factor in the development of United Nations law on the use of force is the practice of states within and outside of the Organization. This state practice consists of the conclusion of treaties [the North Atlantic Treaty Organization Treaty (NATO) and several non-aggression pacts, for example] and claims and counterclaims in diplomatic intercourse regarding the content of applicable norms. Some contend that votes in United Nations organs constitute state practice, although this

---

[37] See UN Charter Article 97.

[38] UN Charter Article 98.

[39] Hammarskjold (xxxx), quoted in Elarably 1987.

proposition is controversial. At a minimum these votes may constitute evidence of *opinion juris* that is, that states are engaging in the practice under a sense of legal obligation.

3.2.5 Articles 2(4) and 51

The visionaries who drafted the UN Charter undoubtedly sought to obviate the use of the word "war" in the Charter. However, because the terms "aggression" and "war"—emphasized in the Covenant of the League of Nations—had caused problems of interpretation for the League in practice, it is apparent that the drafters of the United Nations Charter decided on a different approach. Article 2, para 4 of the Charter provides that:

> All Members shall refrain in their international relations from the threat or use of force against the territorial integrity or political independence of any state, or in any other manner inconsistent with the Purposes of the United Nations.

This notwithstanding, the overall approach was not uniformly followed elsewhere in the Charter, and this has raised confusion as to the proper interpretation and application of Article 2(4). For example, Article I proclaims the primary purpose to be the maintenance of international peace and security by taking effective collective measures to prevent or remove "threats to the peace" and to suppress 'acts of aggression or other breaches of the peace.' It does not use the words "force" or "threat of force". The application of the qualifying words "in conformity with the principles of justice and international law" to the Organization's first purpose, moreover, has led some to claim that, as long as law and justice are served, recourse to force may be justified. Similarly, the principle of self-determination in Article 1 (2) has been cited to support the contention that force can be used on behalf of "wars of national liberation". Also, as discussed later, the Security Council is directed:

> to determine the existence of any threat to the peace, breach of the peace, or act of aggression' and to make recommendations or decide on measures to restore international peace and security. This means that there is "no necessary identity between what is legally prohibited by Article 2(4) and what the Council seeks to control in the discharge of its responsibilities."[40]

Those who favour limited constraints on the right of states to use armed force have relied on textual analysis to interpret Article 2(4) as authorizing a number of exceptions to that article's prohibition against the "threat or use of force." Oscar Schachter observes:

> A more basic question of interpretation (than that concerning the meaning of "force' and "threat of force-] is presented by the peculiar structure of the article. it is generally assumed

---

[40] Goodrich et al. 1969, p. 46 [hereinafter Goodrich].

that the prohibition was intended to preclude all use of force except that allowed as self-defense or authorized by the Security Council under chapter VII of the Charter. Yet the article was not drafted that way. The last twenty-three words contain qualifications. The article requires states to refrain from force or threat of force only when that is 'against the territorial integrity or political independence of any state" or "inconsistent with the purposes Of the United Nations.' If these words are not redundant, they must qualify the all-inclusive prohibition against force. Just how far they do qualify the prohibition is difficult to determine from a textual analysis alone.[41]

Some have claimed that these words greatly qualify the prohibition against force. The qualifying language, for example, would allow the use of force solely to vindicate or secure a legal right, i.e., an exercise in self-help. As Schachter has observed, this argument has enjoyed little governmental or scholarly support.

Similar textual arguments have been made to support the use of force by states to take territory they consider rightfully theirs (e.g., Argentina and the Malvinas/Falkland Islands, Iraq and Kuwait), or a right of "humanitarian intervention," and the right to self-determination (including, in its most recent manifestation, a claim of right to overthrow a repressive or tyrannical government, as well as to achieve freedom from foreign domination). Schachter examined each of these arguments in detail in recent writings[42] and has concluded:

> Many governments attach importance to the principle that any forcible incursion into the territory of another state is a derogation of that state's territorial sovereignty and political independence, irrespective of the motive for such intervention or its long-term consequences. Accordingly, they tend to hold to the sweeping Article 2(4) prohibition against the use or threat of force, except where self-defense or Security Council enforcement action is involved.[43]

In addition to these textual arguments, commentators have argued that recent developments in international relations have significantly altered the conditions on which the restrictive rules were based.[44] The recent developments most often cited are widespread violations of Article 2(4)[45] and the failure of the United Nations system of collective security.[46] Widespread violations of Article 2(4), some contend, constitute state practice that has superseded the Charter provision and its customary law analogue. This argument has several flaws. First, there are relatively few blatant violations of Article 2(4). Moreover, almost every clearly illegal use of armed force has been condemned by large numbers of states and by the political organs of the United Nations. Most important, this argument has not been accepted by states. On the contrary, many states, including the United States, have taken the legal position that Article 2(4) is a peremptory norm (*jus cogens*). The argument

---

[41] Schachter 1984.

[42] *Ibid*. Also see Schachter 1986, p. 113 [hereinafter Schachter, Use of Force); and, Schachter 1989 , p. 259.

[43] Schachter, Use of Force *supra* note 84, at p. 1632.

[44] As reported in Schachter, Use of Force, *supra* note 84, at pp. 113, 124–125.

[45] See Franck 1970, p. 809, Rostow 1985, p. 286.

[46] Thomas and Thomas 1956, p. 209.

also was expressly rejected by the International Court of justice in *Nicaragua v. United States.*[47] There, the Court stated:

> It is not to be expected that in the practice of States the application of the rules in question should have been perfect, in the sense that States should have refrained, with complete consistency, from the use of force or from intervention in each other's internal affairs. The Court does not consider that, for a rule to be established as customary, the corresponding practice must be in absolutely rigorous conformity with the rule. In order to deduce the existence of customary rules, the Court deems it sufficient that the conduct of States should, in general, be consistent with such rules, and that instances of State conduct inconsistent with a given rule should generally have been treated as breaches of that rule, not as indications of the recognition of a new rule. If a State acts in a way prima facie incompatible with a recognition of a new rule, but defends its conduct by appealing to exceptions or justifications contained within the rule itself, then whether or not the State's conduct is in fact justifiable on that basis, the significance of that attitude is to confirm rather than to weaken the rule.[48]

As to the failure of the United Nations collective security system, the legislative history of Article 2(4) lends no support to the thesis that the effective functioning of a system of collective security is regarded as a condition of the continuing validity of the article's severe constraints on the use of force.[49] Moreover, recent developments may have given new life to the Charter's concept of collective security.

A variant of this changed circumstances argument relies on the concept of reciprocity—widespread violations of law by some states should release others from the obligation to comply, an argument by analogy to the right of one treaty party to suspend the treaty between it and a violator for a violation especially affecting that party.[50] Several problems are inherent in this argument. First, as has been discussed, Article 2(4), in addition to being a treaty provision, is widely regarded as an archetypical peremptory norm from which no derogation is permitted. Second, and significantly, the reciprocity thesis finds no support in state practice; no state has cited another state's violation of Article 2(4) as a basis to suspend that provision's obligations in relations between it and the violator. Third, the Charter permits a variety of responses to a breach of Article 2(4): the right to self-defence in the event of an armed attack; recourse to the Security Council or the General Assembly; and, if a jurisdictional basis is available under its statute, to the International Court of justice.[51]

The justification for recourse to armed force most often cited by states is, of course, the right to self-defence. Article 51 provides:

---

[47] Military and Paramilitary Activities in and against Nicaragua (*Nicar. v. US*) 1986 ICJ Rep. 14, 98 (Judgment of June 27).

[48] *Id.* at p. 98.

[49] See, e.g., Noncic 1970, pp. 76–77 [guarantees of territorial integrity in art. 2(4) were not intended to limit the article's broad prohibition of self-help measures], cited in Schachter, Use of Force, *supra* note 84, at pp. 125–126 n. 56.

[50] See Address by Ambassador Kirkpatrick 1984, pp. 59, 67.

[51] See Schachter, Use of force, *supra* note 84, at pp. 129–130.

Nothing in the present Charter shall impair the inherent right of individual or collective self-defence if an armed attack occurs against a Member of the United Nations, until the Security Council has taken measures necessary to maintain international peace and security. Measures taken by Members in the exercise of this right of self-defence shall be immediately reported to the Security Council and shall not in any way affect the authority and responsibility of the Security Council under the present chapter to take at any time such action as it deems necessary in order to maintain or restore international peace and security.

This provision applies to self defence which is usually exercised in response to an attack in order to repulse that attack. It is also noted that the United Nations Security Council has the power to take measures to maintain or restore peace.

3.2.6 The Law of the Sea Analogy

Since the Outer Space Treaty disallows national appropriation of outer space by means of use or occupation or by any other means, and prescribes that States shall carry out activities in the exploration and use of outer space, including the Moon and other celestial bodies, *in accordance with international law*, including the Charter of the United Nations, the law pertaining to the high seas[52] is directly relevant as being an example of international law that could apply in instances where life may be discovered in outer space. This is because the essence of the freedom of the high seas lies in the fact that no State may acquire sovereignty over parts of the high seas.[53] However, the 1951 decision of the International Court of Justice, handed down in the *Anglo-Norwegian Fisheries* case[54] recognized the doctrines of recognition, acquiescence and prescription where, by long usage acknowledged by other States, certain areas of the high seas adjoining territorial waters of a State may be considered subject to that State's sovereignty. It is not clear as to how this principle could apply in space law as no State can claim a celestial body or portion of outer space as adjoining its territory. Therefore, one can only assume that the pristine principle enunciated in the Outer Space Treaty, that no State can claim sovereignty over outer space, will still prevail.

Another area in which space law and the law of the sea have a common denominator is jurisdiction. Article VI of the *Outer Space Treaty* provides in part that State Parties to the Treaty shall bear international responsibility for national activities in outer space, whether such activities are carried out by governmental agencies or non-governmental agencies. This provision clearly introduces the notion of strict liability *erga omnes* to the application of the *jus cogens* principle

---

[52] The high seas were defined in Article 1 of the 1958 Geneva Convention on the High Seas as comprising all parts of the sea that were not included in the territorial sea or in the internal waters of a State.

[53] Article 2 of the 1958 Geneva Convention on the High seas, and Article 89 of the Convention on the Law of the Sea, 1982.

[54] ICJ Reports 1951, p 116. 18 ILR p. 86.

relating to outer space activities of States and could be considered applicable in instances where States hold out to the international community as providers of technology achieved and used by them in outer space, which is used for purposes of air navigation. Article VI further requires that the activities of non-governmental entities in outer space shall require authorization and continuing supervision by the appropriate State Party to the Treaty, thus ensuring that the State whose nationality the entity bears would be vicariously answerable for the activities of that organization, thereby imputing liability to the State concerned.

Article VII makes a State Party internationally liable to another State Party for damage caused by a space object launched by that State. The *Registration Convention* of 1974[55] in Article II(1) requires a launching State of a space object that is launched into earth orbit or beyond, to register such space object by means of an entry in an appropriate registry which it shall maintains and inform the Secretary General of the United Nations of the establishment of such a registry. This provision ensures that the international community is kept aware of which State is responsible for which space object and enables the United Nations to observe outer space activities of States. Article VI of the Convention makes it an obligation of all State Parties, including those that possess space monitoring and tracking facilities, to render assistance in identifying a space object which causes damage to other space objects or persons.

Analogically, the foundation of the maintenance of order on the high seas has rested squarely on the nationality of the ship and the attendant jurisdiction of the flag State over the ship. Therefore responsibility for an act performed on the high seas would rest on the flag State of the ship, as much as corresponding responsibility in outer space would devolve upon the State of registry of the space craft. Of direct bearing on this issue is piracy where the law of the sea has clear laws that could be relevant to an instance where life in outer space is discovered by a non State party.

Criminal conduct is an area where the principle of international law applicable to the High seas lend themselves as a useful analogy to space law. Of course, the offence of piracy cannot be committed by astronauts who are sent to outer space in spacecraft belonging to a State. The offence has to be committed for private ends by persons in a private ship or craft. The offence of piracy in the high seas would nonetheless apply as an analogy to a similar offence committed by private individuals in outer space who do not represent a State as official crew members. This would cover the improbably but nonetheless possible events of the future such as a mutiny on board a commercial spacecraft carrying passengers (which is an analogy derived from shipping law). Piracy in outer space may also occur in instances where personnel of a space craft could act on the orders of a recognized government which is in gross breach of international law and which show a criminal disregard for human life.

---

[55] *Convention on Registration of Objects Launched into Outer Space*, Adopted by the General Assembly of the United Nations, New York, 12 November 1974, 1023 UNTS 15.

The offence of piracy at sea and its consequences were succinctly defined by Judge Moore in his dissenting judgment in the *Lotus* Case.[56]

> In the case of what is known as piracy by law of nations, there has been conceded a universal jurisdiction, under which the person charged with the offence may be tried and punished by any nation into whose jurisdiction he may come. I say "piracy by law of nations", because the municipal laws of many States denominate and punish as "piracy" numerous acts which do not constitute piracy by law of nations, and which therefore are not of universal cognizance, so as to be punishable by all nations. Piracy by law of nations, in its jurisdictional aspects, is *sui generis*. Though statutes may provide for its punishment, it is an offence against the law of nations; and as the scene of the pirate's operations is the high seas, which it is not the right or duty of any nation to police, he is denied the protection of the flag which he may carry, and is treated as an outlaw, as the enemy of all mankind - *hostis humani generis* - whom any nation may in the interest of all capture and punish.[57]

Article 15 of the Convention on the High seas of 1958 defines the offence of piracy as the following: Any illegal acts of violence, detention or any act of depredation, committed for private ends by the crew or the passengers of a private ship or a private aircraft, and directed: on the high seas, against another ship or aircraft, or against persons or property on board such ship or aircraft; against a ship, aircraft, persons or property in a place outside the jurisdiction of any State. any act of voluntary participation in the operation of a ship or of an aircraft with knowledge of facts making it a pirate ship or aircraft; any act of inciting or of intentionally facilitating an act described in sub-paragraph (1) or sub-paragraph (2) of this Article. *Mutatis mutandis*, this provision would serve well as an analogy and persuasive authority in the event a similar offence committed in outer space or celestial body is examined by a competent court of any jurisdiction.

The Convention restricts the application of the offence of piracy to acts on the high seas or "any place outside the territorial jurisdiction of any state" which means essentially *in contextu* an island or "*terra nullius*" (no man's land). The latter is a fitting analogy for outer space or celestial body which is outside the jurisdiction of any State.

Article 19 of the Convention on the High seas provides for remedial action and grants the right to any State on the high seas, or in any other place outside the jurisdiction of that State, to seize a pirate ship or aircraft, or a ship taken by piracy and under the control of pirates, and arrest the persons and seize the property on board. The courts of the State which carried out the seizure may decide upon the penalties to be imposed, and may also determine the action to be taken with regard to the ships, aircraft or property, subject to the rights of third parties acting in good faith.

The significance of the provisions of the Convention with regard to the offence of piracy, when related to similar offences which may be committed in outer space, lies in the universal application of the various limits of the offence itself. For example the offence of piracy includes acts of violence, detention or any act of

---

[56] PCIJ, ser A, no. 10 (1927).

[57] *Id.* At p. 70.

depredation. This effectively brings to bear its relevance to instances such as the take over, detention or illegal deviation of a spacecraft and therefore is quite useful as paradigm legislation for space law.

With regard to principles of tortious liability, both space law and shipping law share a common thread of international responsibility of States. Space law, by virtue of the principle that a State of registry retains jurisdiction in its space object, imputes territoriality to a State in relation to its space craft, space station or other space object. Therefore, any act committed by one State or individual representing a State which would adversely affect such space object would be comparable with an instance where the act of one State affects another in its territorial waters.

Principles of State responsibility for negligent acts leading to reparation are concomitants of substantive public international law. International responsibility relates both to breaches of treaty provisions to which States are bound and also to other infringements of legal duty. Brownlie makes bold to assert that there is no harm in using the term "international tort" to describe the breach of a duty which results in a loss to another State.[58] Judge Huber in the *Spanish Zone of Morocco Claims* case observed:

> Responsibility is the necessary corollary of a right. All rights of an international character involve international responsibility. If the obligation in question is not met, responsibility entails the duty to make reparation.[59]

In the celebrated *Corfu Channel* case[60] which involved the liability of Albania for mine-layering in her international waters which caused space damage to British mine sweepers, the Court held:

> These grave omissions involve the international responsibility of Albania. The Court therefore reaches the conclusion that Albania is responsible under international law for the explosions which occurred ... and for the damage and loss of human life which resulted from them.[61]

The above principle will incontrovertibly apply equally to space law for two reasons: both space law and the law related to the Sea are grounded in the principles of public international law; and principles of State responsibility are universal and cannot be applied on different bases with regard to species of international law.

The responsibility of States for the negligent acts of their agents in outer space would be based on the concept of objective responsibility i.e.: responsibility of States for those acts committed by their officials or its organs and which acts such officials are bound to perform in the course of their duties. The elements of *faute* or fault on the part of the officials is irrelevant. Tortious liability of the State concerned and responsibility therefor would also entail in instances where the officials or State agencies act beyond their competence. The only necessary element for the

---

[58] Brownlie 1990 at p. 434.

[59] RIAA ii 615, at p. 641.

[60] ICJ Reports (1949) at p. 23.

[61] *Ibid.*

imposition of State responsibility is the establishment of the fact that such persons acted with authority derived from the State concerned when the tortious act was performed.

Although the nature of outer space and the high seas and the uses they are put to, are different the underlying principles with regard to the conduct or persons in the two geographic areas are similar, if not identical. Treaty provisions with regard to the use of both are certainly comparable, and to this extent space law could alternate principles from the law of the sea when necessary.

3.2.7 Conduct of the Space Tourist

Space tourism is any commercial activity that offers customers direct or indirect experience with space travel.[62] In Space Law, there is no such being as a "person" in outer space. There are only astronauts and personnel.[63] The 1967 Outer Space Treaty stipulates that State parties to the Treaty shall regard astronauts as envoys of mankind in outer space and shall render to them all possible assistance in the event of accident, distress or emergency landing on the territory of another State party or on the high seas.[64] The provision also requires State parties to return astronauts under the above circumstances safely and promptly to the State of registry of their space vehicle.[65]

The Treaty provision is a reproduction *verbatim* of para 9 of United Nations General Assembly Resolution XVIII of 1962.[66] Although initially, the world's "envoys of mankind" seemingly created some apprehension in the international community as to whether such phraseology connoted diplomatic immunity to astronauts, Bin Cheng clears up this ambivalence by concluding that it was only a figure of speech which has not been repeated in any United Nation's documentation yet.[67] The perceived inadequacy of definitive identification at international law of an astronaut and his conduct in outer space leaves one with the basic premise that international law is incontrovertibly applicable to outer space activities and outer space, including the moon and other celestial bodies, which are totally independent

---

[62] Hobe and Cloppenburg 2004 at p. 377 cited in Chatzipanagiotis 2011 at p. 1.

[63] Treaty on Principles Governing the Activities of States in the Exploration and Use of Outer Space, Including the Moon and Other Celestial Bodies, Opened for Signature at Moscow, London and Washington on 27 January 1967, 610VNTS 205. It must be noted that the first "space tourist" Denis Tito was called a space tourist for purposes of public reference by the media. He was called a "guest cosmonaut" by the Russians and an amateur astronaut by the Americans. See http://www. spacedaily.com/news/011206133411.3i4zwq28.html.

[64] *Id.* Article V.

[65] *Ibid.*

[66] UNGA Resolution 1962 (XVIII) Declaration of Legal Principles Governing the Activities of States in the Exploration and Use of Outer Space.

[67] Bin 1997 at pp. 259 and 460.

of appropriation by States or individuals. This in turn leaves one with the inevitable conclusion that outer space would be analogous at international law to the high seas.

A commercial space traveller who pays an operator to be carried into outer space is similar to a commercial air traveller and, socially, the former's conduct and social behaviour should accord with the same principles as applicable to the latter. In this context the Tokyo Convention could be a good analogy, the principles of which can be adapted to suit commercial space travel. The Tokyo Convention, which was signed on 14 September 1963, entered into force on 4 December 1969 and has 185 State Parties which have signed or ratified it. According to the Convention, the aircraft commander has the power to take measures and restrain a person even if his act did not amount to jeopardising the safety of the aircraft or the person or the property therein. This provision directly leads to an analogous situation of an offensive act perpetrated by an unruly person. If for example the conduct of two or more persons, while on board the aircraft, appear to be leading to the commission of some illegal act (the conduct of a drunken sporting team on board would be a good example), the commander can restrain them on the suspicion that they are conspiring to commit an act which may jeopardise good order and discipline on Board. This seems to be totally consistent with the general thrust of the Convention and its principal objective which is to assure the maintenance of safety and good order "on board" the aircraft.

The aircraft commander, in discharging his duties according to the Convention, can require or authorize the assistance of the crew and request the assistance of persons for that purpose. Even persons and crew members are authorized under Article 6(2) to take reasonable preventive measures without any authorization from the aircraft commander whenever they have reasonable grounds to believe that such action is immediately necessary for safety reasons. Although this clause is wide enough to give powers to other people beside the aircraft commander in order to tighten the measures that leads to the thwarting of acts of unlawful interference with civil aviation, some delegates at the Tokyo Conference attacked this approach on the ground that persons normally would not be qualified to determine whether a particular act jeopardized the safety of the aircraft or persons and property therein, and, for this reason, it was unwise to give this authority to persons.[68] However, this argument was rejected "on the ground that this provision contemplated an emergency type of situation on which the danger of the aircraft or persons and property on board was clearly present, and in fact no special technical knowledge would be required to recognize the peril.

The powers entrusted to the commander in order to suppress any unlawful act that threatens the safety of the aircraft go as far as requiring the disembarking of any person in the territory of any State in which he lands and delivering him to its competent authorities.[69] The State is under an obligation to allow the

---

[68] *Ibid.*

[69] Article 8 of the Tokyo Convention.

disembarkation and to take delivery of the person so apprehended by the aircraft commander, but such custody may only be continued for such time as is reasonably necessary to enable the criminal extradition proceedings (if any) to be instituted. In the meantime the State of landing should make a preliminary enquiry into the facts and notify the State of registration of the aircraft.[70]

In any event, the commander as well as the crew members and persons are given immunity from suits by the alleged offender against whom they acted. Article 10 expressly provides:

> Neither the aircraft commander, any member of the crew, any person, the owner or operator of the aircraft, nor the person on whose behalf the flight is performed shall be held responsible in any proceedings on account of the treatment undergone by the person against whom the actions were taken.

This protection was given to the aircraft commander and other persons in order to encourage them to fight the wrongful acts contemplated by the Convention.

3.2.8 Concluding Remarks

The *Moon Agreement* of 1979 provides that in the exploration and use of the moon, States Parties shall take measures *inter alia* to avoid harmfully affecting the environment of the earth through the introduction of extra terrestrial matter or otherwise.[71] At the same time, the Moon agreement, which applies to other celestial bodies as well, in Article 6 provides that there shall be freedom of scientific investigation on the moon by all States Parties without discrimination of any kind, on the basis of equality and in accordance with international law. The same provision allows a State Party which finds minerals and other substances to collect and take custody of such material and even conduct scientific experiments on the samples. They are required to have regard to requests from other State Parties for the use of such material and also given the discretion to exchange scientific information in this context. In this context it must be noted that at the frontiers of this issue are the astronauts,[72] who are by treaty designated as envoys of mankind in outer space, casting on them the responsibility of adhering to applicable treaty provisions on behalf of their States.

In the field of international space law, two clearly connected terms have been used: liability and responsibility. Although "responsibility" has not been cohesively interpreted in any legal treaty relating to outer space, "liability" occurs in the *Convention on International Liability for Damage Caused by Space Objects*, March 29 1972 (Liability Convention) and is sufficiently clear therein. This, however, does not mean that State responsibility is not relevant to the obligations of States law as,

---

[70] See Articles 12 and 13 of the Tokyo Convention.

[71] Article 7.

[72] Outer Space Treaty, *supra* note 106, Article V.

in international relations, the invasion of a right or other legal interest of one subject of the law by another inevitably creates legal responsibility.

At present, one can regard responsibility as a general principle of international law, a concomitant of substantive rules and of the supposition that acts and omissions may be categorized as illegal by reference to the rules establishing rights and duties. Shortly, the law of responsibility is concerned with the incidence and consequence of illegal acts, and particularly the payment of compensation for loss caused. Therefore, As discussed, both treaty law and general principles of international law on the subject of space law make the two elements of liability and responsibility a means to an end—that of awarding compensation to an aggrieved State or other subject under the law. In view of the many legal issues that may arise, the primary purpose of a regulatory body which sets standards on State liability would be to carefully consider the subtleties of responsibility and liability and explore their consequences on States and others involved as they apply to the overall concept of the status of a State as a user of space technology which may cause harm or injury to the latter.

Finally, we have to be mindful of a few fundamental truths. First, if we come across any form of life in outer space it will be the concern of all humankind. Second, any treatment of such life, irrespective of the fact that it is found in outer space, should be according to the principles of international law and the United Nations Charter. Within these parameters, yes, we could send more probes to investigate further. Yes, we could even put up a "do not disturb sign". But whatever we do, we are bound by the principles of responsibility and international accountability to treat life in outer space with the same dignity that should be accorded to life on Earth and to strictly adhere to the principles of international law and the United Nations Charter in ensuring peace in the process.

References

Arnold K, Palmer, R (1952) The coming of the saucers, p 3

Bin C (1997) Studies in international space law. Clarendon Press, Oxford, pp 259, 460

Brownlie I (1990) Principles of public international law, 4th edn. Clarendon Press, Oxford, p 434

Chatzipanagiotis M (2011) The legal status of space tourists in the framework of commercial suborbital flights. Carl Heymanns Verlag, Cologne, p 1

Elarably (1987) The office of the secretary general and the maintenance of international peace and security. In United Nations and the maintenance of international peace and security, vol 177, pp 184–95

Franck T (1970) Who killed Article 2(4)? On changing norms governing the use of force by states. AJIL 64:809

Goodrich LM, Hambro E, Simons A (1969) Charter of the United Nations, p 46

Hammarskjold D The international civil servant in law and in fact. United Nations Press Release SG/1035, p 17

Hobe S, Cloppenburg J (2004) Towards a new aerospace convention. In: Selected legal issues of space tourism, IISL Proceedings, vol 47, p 377

Hynek JA (1972) The UFO experience. H. Regnery Co., Chicago, p 12

Kirkpatrick J (1984) Law and reciprocity. In: Proceedings of the 78th annual meeting of the American society of international law, vol 59, p 67 (reprinted)

Noncic D (1970) The problem of sovereignty in the Charter and in the practice of the United Nations 72:76–77

Rosenstock R (1971) The declaration of principles of international law concerning friendly relations: a survey. AJIL 65:713, 714

Rostow E (1985) The legality of the international use of force by and from states. Yale J Int L 10:286

Ruwantissa A (2013) The regulation of air transport—the slumbering sentinels. Springer, New York

Schachter O (1984) The right of states to use armed force. Mich L R 82:1620, 1625

Schachter O (1986) In defense of international rules on the use of force. U Chi L Rev 53:113

Schachter O (1989) Self-defense and the rule of law. AJIL 83:259

Smith, MS (2006) The vision for expanding the envelope for space law debates. J Space Law 32 (1):217, 221

Thomas AV, Thomas AJ Jr (1956) Non-intervention. Blackwell, New York, p 209

Chapter 4
ICAO for Commercial Space Travel?

4.1 Background Activity

The Conference on the Regulation of Emerging Modes of Aerospace Transportation (REMAT), held on 24–25 May 2013 in Montreal, discussed among other issues the subject of ICAO for space, with special focus on commercial space transportation. REMAT recognized ICAO as having a structured framework of Standards and Recommended Practices (SARPs) on various areas of aviation such as air traffic management, personnel licensing, rules of the air and airport planning, with a view to determining whether these SARPs could be extended to encompass principles that could be applied to the regulation of commercial space transport. Organized by McGill's Institute of Air and Space Law, the International Civil Aviation Organization, and the International Association for Advancement of Space Safety, in ccollaboration with the International Foundation for Aviation, Aerospace & Development, the Conference aimed at assessing the current situation and future plans for aerospace transportation; critically examining and identifying precisely the regulatory challenges to operation of aerospace vehicles; and suggesting viable policy and regulatory steps (mechanisms) that may be considered by States and other stakeholders to facilitate aerospace transportation, thus ensuring the safety of global aviation.

REMAT had some interesting presentations on the scientific issues of commercial space transportation on the Rocket Plane, Virgin Galactic and other aero-spacecraft as well as the conditions that would apply to space tourists during sub orbital flights. Interesting questions were raised; concerning such issues as what type of insurance regime would apply to commercial space travel and space tourists; whether insurers would insist on such travellers using outfits worn by astronauts; and whether space tourists would need intense training such as the training which astronauts undergo prior to space flight.

Although there was an impressive ICAO presence among the presenters, who generally discussed the fact that existing SARPs of the Annexes to the Convention

© The Author(s) 2015
R. Abeyratne, *Regulation of Commercial Space Transport*,
SpringerBriefs in Law, DOI 10.1007/978-3-319-12925-9_4

on International Civil Aviation (Chicago Convention) may be stretched to apply to commercial space travel, no specifics were discussed. There was only one ICAO lawyer who was relegated to the final Panel of discussion and he was asked two mundane questions on aviation. The author expected some discussion on the modalities of transition of ICAO as a specialized agency of the United Nations on aviation towards becoming an agency for space travel as well, and asked the Conference the following questions: can we just simplistically say that this issue could be addressed just by extending ICAO's "mandate" (pointing out that ICAO does not have a mandate but aims and objectives under the Chicago Convention which were purely on aviation)?; if not, how could this be done?; what about Article 37 of the Chicago Convention which exclusively dealt with SARPs on aviation? Are we just discussing ICAO's role in ensuring the safety of aviation when it came to sub orbital flights or are we talking about ICAO as an Organization addressing near space and outer space issues? No answers came forth.

The above notwithstanding, REMAT came across as a proactive first step towards further discussion and if one were to take just one consensual point from the conference it was that there was much to be discussed among all key stake-holders before such an issue could be brought before the ICAO Council.

4.2 Amending the Chicago Convention

Assad Kotaite, a former President of the ICAO Council, in his Foreword to the Study which was referred to in the Preface has said that merely by changing ICAO's "mandate" the regulation of commercial space transport could be taken over by ICAO. What Kotaite probably meant (although he did not explicitly say so) was that Article 44 of the Chicago Convention which spells out the aims and objectives of ICAO could have an Article 44 *bis* that added on to the list of Annexes to the Convention an Annex on commercial space transport. However, it is not as simple as that. Article 44 would have to be amended by including in ICAO's aims and objectives one that would cover meeting the needs of the people of the world for safe, regular, economic and efficient commercial space transport. In addition, as discussed above, under safety of commercial space transport, and unless a new Convention for commercial space transport is adopted, almost the entirety of the Chicago Convention may have to be adapted by having alternate provisions for safety. Firstly, amending the Chicago Convention could prove to be a tedious process. Article 94 of the Chicago Convention, which addresses the issue of amendment of the Convention states:

"Article 94

Amendment of Convention

(a) Any proposed amendment to this Convention must be approved by a two-thirds vote of the Assembly and shall then come into force in respect of States which have ratified such amendment when ratified by the number of contracting States specified by the Assembly. The number so specified shall not be less than two thirds of the total number of contracting States.

(b) If in its opinion the amendment is of such a nature as to justify this course, the Assembly in its resolution recommending adoption may provide that any State which has not ratified within a specified period after the amendment has come into force shall thereupon cease to be a member of the Organization and a party to the Convention".

As one can readily note, there are some key considerations *in limine*. Firstly, an amendment to the Convention must be approved by two thirds vote of the members present and voting at the Assembly. Thereafter it has to be ratified by a number of States as specified by the Assembly and such amendment will be applicable to those States that voted for it. The Assembly cannot specify for ratification any number less than two thirds of the membership of ICAO (which at the present time would be 127 States).[1] Amendment to Article 50(a), signed at Montreal on 21 June 1961; Amendment to Article 48(a), signed at Rome on 15 September 1962; Amendment to Article 50(a), signed at New York on 12 March 1971; Amendment to Article 56, signed at Vienna on 7 July 1971; Amendment to Article 50(a), signed at Montreal on 16 October 1974; Amendment (final paragraph referring to the authentic Russian text), signed at Montreal on 30 September 1977; Amendment (Article 83 *bis*), signed at Montreal on 6 October 1980; Amendment (Article 3 *bis*), signed at Montreal on 10 May 1984; Amendment to Article 56, signed at Montreal on 6 October 1989; Amendment to Article 50(a), signed at Montreal on 26 October 1990; Amendment (final paragraph referring to the authentic Arabic text), signed at Montreal on 29 September 1995; Amendment (final paragraph referring to the authentic Chinese text), signed at Montreal on 1 October 1998. Of these only the last two amendments have not come into force. Would such ratification practically attain fruition? And, more ominously, the Assembly Resolution recommending adoption of an amendment could provide that any State that does not ratify within a specified time would cease to be a member of ICAO and a party to the Convention.

Would States jeopardize their membership of ICAO for issues of space regulation? Would they harken to the prospect of mixing aeronautics with space transport in the same provision of the Chicago Convention or an Annex thereto?

If space transportation regulations were to be introduced within the ICAO umbrella as suggested, in practicality, the initial work would fall upon the Legal Committee of the Council of ICAO. The ICAO Assembly, at its 7th Session (Brighton 16 June–6 July 1953) adopted Resolution A7-5 (Revised Constitution of the Legal Committee) which resolved that the Legal Committee shall be a permanent Committee of ICAO constituted by the Assembly and responsible to the Council except as otherwise specified. The duties of the Committee are to advise

---

[1] For the past 67 years of its existence, the Chicago Convention has been amended 15 times as follows:. Amendment (Article 93 *bis*), signed at Montreal on 27 May 1947; Amendment to Article 45, signed at Montreal on 14 June 1954; Amendment to Articles 48(a), 49(e), and 61, signed at Montreal on 14 June 1954.

the Council on matters relating to the interpretation and amendment of the Chicago Convention referred to it by the Council; to study and make recommendations on such other matters relating to public international air law as may be referred to it by the Council or Assembly to study problems relating to private air law affecting international civil aviation as directed by the Council or Assembly of ICAO and to make recommendations on participation of non-Contracting States and other international Organizations at meetings of ICAO. As to how the Legal Committee —a body of experts on air law—would be constituted to cope with space transportation issues would be another matter for consideration.

ICAO has another fundamental problem. An aircraft has been explicitly defined.[2] Although there is no accepted definition of a spacecraft, it could be taken to be a manned or unmanned vehicle designed to orbit the earth or travel to celestial objects for the purpose of research, exploration. The Study[3] states that under the definition of "aircraft" an aerospace vehicle launched by a rocket would not be considered an aircraft on the ascent phase of its flight, but might well come within the definition during its descent phase as it glides towards its destination.[4] There is also some confusion as to what a space object is which is governed by the principles of space law as none of the 5 Space Law Conventions mention such an object. On the other hand, there could be an "aerospace vehicle" which is a hybrid aerospace object that is capable of achieving lift and flying in airspace as well as in outer space. How would the Chicago Convention or any of its Annexes handle such a complex array of "objects"?

The Study makes a clear distinction:

> But it is unclear whether a commercial aerospace vehicle constitutes a civil aircraft... During the ballistic portions of its flight, while not supported by the reactions of the air, a spacecraft would not fall under this peculiar definition. However, upon descent, after it has re-entered the earth's atmosphere and as it is gliding on its return to Earth, it could be considered an aircraft in flight, and therefore subject to the Chicago Convention.[5]

This only makes confusion worse confounded. If an aerospace vehicle (or spacecraft) can only be identified as an aircraft in its descent phase when it is guiding towards its destination on the ground, would it be prudent to amend drastically the Chicago Convention and its Annexes as relevant to apply to all phases of a spacecraft? For instance, How would flight information regions (FIRs) which are governed by the principles contained in Annex 11 to the Chicago Convention on air traffic services be affected? Would they be relevant only at the descent phase of the spacecraft or would the entirety of Annex 11 be extended to cover communications regarding navigation in space? All aircraft fly in accordance with either instrument flight rules (IFR) or visual flight rules (VFR). Under IFR, the

---

[2] An aircraft is any machine that can derive support in the atmosphere from the reactions of the air other than the reactions of the air against the earth's surface. See ICAO Annex 1, Annex 6 Part I.

[3] ICAO for Space?, *supra*, note 14.

[4] *Id.* 59.

[5] *Id.* 61.

aircraft fly from one radio aid to the next or by reference to self-contained airborne navigation equipment from which the pilot can determine the aircraft's position at all times. Would this practice be extended to outer space transportation?

Annex 3 to the Chicago Convention provides in Standard 2.1.1. that the objective of meteorological service for international air navigation shall be to contribute towards the safety, regularity and efficiency of international air navigation. This objective is achieved by supplying the following users: operators, flight crew members, air traffic services units, search and rescue services units, airport managements and others concerned with the conduct or development of international air navigation, with the meteorological information necessary for the performance of their respective functions.[6] How would this Annex accommodate meteorological conditions in outer space?

Annex 2 to the Chicago Convention prescribes rules of the air. The Annex requires that an aircraft must be flown in accordance with the general rules and either the visual flight rules (VFR) or the instrument flight rules (IFR). Flight in accordance with visual flight rules is permitted if a flight crew is able to remain clear of clouds by a distance of at least 1,500 m horizontally and at least 300 m (1,000 ft) vertically and to maintain a forward visibility of at least 8 km. For flights in some portions of the airspace and at low altitudes, and for helicopters, the requirements are less stringent. An aircraft cannot be flown under VFR at night or above 6,100 m (20,000 ft) except by special permission. Balloons are classified as aircraft, but unmanned free balloons can be flown only under specified conditions detailed in the Annex. Would there be an extension to this Annex on rules of outer space? Or would there be a new Annex to the Chicago Convention on this subject?

The ICAO Council, in adopting Annex 2 in April 1948 and subsequently in November 1951 when Amendment 1 to the Annex was adopted, resolved that the Annex constitutes *rules* relating to the flight and manoeuvre of aircraft within the meaning of Article 12 of the Convention. Therefore, the Council explicitly recognized that the rules in the Annex applied to the manoeuvre and operation of aircraft without exception. Annex 2, in its Foreword, states that the Standards in the Annex, together with the Standards and Recommended Practices of Annex 11, govern the application of the Procedures for Air Navigation Services Rules of the Air and Air Traffic Services, and the Regional Supplementary Procedures. The Regional Supplementary Procedures are subsidiary procedures of regional applicability. It is clear that by this introduction, there is established a distinct disparity between Annex 2 and Annex 11 where the provisions of the former remain unquestionably mandatory, and the provisions of the latter remain subject to Article 38 of the Chicago Convention and capable of being deviated from. However, it is clear that the purpose of Annex 11 is to ensure that flying on international routes is carried out under uniform conditions designed to improve the safety and efficiency of air operation and, therefore, provisions relating to air traffic control services, flight information services, and alerting services of Annex 11 when linked to the

---

[6] Standard 2.1.2.

provisions of Annex 2, have a coercive effect that may in certain circumstances, transcend the parameters set in Article 38 of the Convention. In this context, ICAO may wish to give careful consideration to the adaptability of both Annex 2 and 11 to fit into a regime of commercial air transport.

Another key issue in this context would be accident investigation. Article 26 of the Chicago Convention provides that, in the event of an accident to an aircraft of a contracting State occurring in the territory of another contracting State, and involving death or serious injury, or indicating serious technical defect in the aircraft or air navigation facilities, the State in which the accident occurs will institute an inquiry into the circumstances of the accident, in accordance, so far as its laws permit, with the procedure which may be recommended by ICAO. Article 26 goes on to say that the State in which the aircraft is registered shall be given the opportunity to appoint observers to be present at the inquiry and the State holding the inquiry shall communicate the report and findings in the matter to that State. Would this principle be extended to cover spacecraft accident investigation under ICAO or the Chicago Convention? These provisions will have to be considered against existing accident provisions in the space law regime, which will be discussed later.

Article 32 of the Chicago Convention provides that the pilot of every aircraft and the other members of the operating crew of every aircraft engaged in international navigation shall be provided with certificates of competency and licenses[7] issued or rendered valid by the State in which the aircraft operates. The provision also states that each ICAO member State reserves the right to refuse to recognize, for the purpose of flight above its own territory, certificates of competency and licences granted to any of its nationals by another member State. Member States of ICAO, at the 21st Session of the ICAO Assembly in Resolution A21-21, which has been alluded to earlier, in Appendix A resolves that certificates of airworthiness and certificates of competency and licenses of the crew of an aircraft issued or rendered valid by the ICAO member State in which the aircraft is registered shall be recognized as valid by the other States for the purpose of flight over their territories, including landings and take offs subject to the provisions of Articles 33 and 32(b) of the Chicago Convention.

Article 33 provides that Certificates of airworthiness and certificates of competency and licenses issued or rendered valid by the contracting State in which the aircraft is registered, shall be recognized as valid by the other contracting States, provided that the requirements under which such certificates or licences were issued or rendered valid are equal to or above the minimum standards which may be established from time to time pursuant to the Convention. Article 32(b) provides, as

---

[7] The expression "licence" used throughout this Annex has the same meaning as the expressions "certificate of competency and license", "license or certificate" and "license" used in the Convention. Similarly the expression "flight crew member" has the same meaning as the expressions "member of the operating crew of an aircraft" and "operating personnel" used in the Convention while the expression "personnel other than flight crew members" includes the expression "mechanical personnel" used in the Convention.

mentioned earlier that each ICAO member State reserves the right to refuse to recognize, for the purpose of flight above its own territory, certificates of competency and licences granted to any of its nationals by another member State. How would corresponding astronauts' licences be incorporated under ICAO or the Chicago regime?

One of the suggestions in the Study alluded to earlier is that:

> …ICAO could promulgate a new Annex on "Space Standards". There is precedent for this as well. Article 37 of the Chicago Convention vests in ICAO the authority to promulgate Standards and Recommended Practices as Annexes to the Convention.[8]

The fundamental flaw in this argument is that the precedent cited pertaining to Article 37 pertains purely to aviation and there could not be "space standards" in an Annex unless these standards relate to the decent phase of a spacecraft as the Study itself suggests. A Convention solely dedicated to international civil aviation cannot include "space standards" pertaining to navigation in outer space.

4.3 Adapting the Annexes to the Chicago Convention

The next issue to consider is, would one composite Annex, say, Annex 20 to the Convention cover all aspects of safety, economics and social needs of people relating to commercial space transport. Or, would there have to be separate Annexes? Let us take the latter scenario with a few examples.

4.3.1 Personnel

Annex 1 contains Standards and Recommended Practices adopted by the International Civil Aviation Organization as the minimum standards for personnel licensing. The Annex is applicable to all applicants for and, on renewal, to all holders of the licences and ratings specified herein. The ICAO Council has decided that, in principle, amendments affecting existing licensing specifications are applicable to all applicants for, and holders of, licences but, in considering their application to existing holders of licences, the assessment, if necessary, by re-examination of the knowledge, experience and proficiency of individual licence holders is left to the discretion of Contracting States.

As long as air travel cannot do without pilots and other air and ground personnel, their competence, skills and training will remain the essential guarantee for efficient and safe operations. Adequate personnel training and licensing also instil confidence between States, leading to international recognition and acceptance of

---

[8] *Supra,* at p. 63.

personnel qualifications and licences and greater trust in aviation on the part of the traveller. Standards and Recommended Practices for the licensing of flight crew members (pilots, flight engineers and flight navigators), air traffic controllers, aeronautical station operators, maintenance technicians and flight dispatchers, are provided by Annex 1 to the Chicago Convention.

Related training manuals provide guidance to States for the scope and depth of training curricula which will ensure that the confidence in safe air navigation, as intended by the Convention and Annex 1, is maintained. These training manuals also provide guidance for the training of other aviation personnel such as aerodrome emergency crews, flight operations officers, radio operators and individuals involved in other related disciplines. In the context of personnel licensing of spacecraft crew, provisions such as Standard 2.5 of the Annex 2 which states that no person whose function is critical to the safety of aviation (safety-sensitive personnel) shall undertake that function while under the influence of any psycho-active substance, by reason of which human performance is impaired, and no such person shall engage in any kind of problematic use of substances, will have to be adapted.. Furthermore, Standard 1.2.7.1 of Annex 1 on personnel licensing provides that holders of licences provided for in the Annex shall not exercise the privileges of their licences and related ratings while under the influence of any psychoactive substance which might render them unable to safely and properly exercise such privileges. This will also have to be adapted.

Then of course comes the issue of cabin attendants in the spacecraft. Would there be any? and if so, the current dilemma facing commercial air transport will haunt commercial space travel as well. There is no doubt that cabin crew form an integral part of commercial aviation, and they should also come under universal training methods and codes of conduct as do the pilots, mechanics, aeronautical engineers and other professionals who are involved with the successful operation of a commercial flight. There is a compelling need for the international aviation community to require a serious study relating to the feasibility of introducing a unified system of rules relating to the conduct of cabin crew, which could *inter alia*, include principles of protection of cabin crew and provide for compensation in case of injury. After all, they are the only ones who deal with the "human factor" of a flight, which could be most unpredictable at the best of times. In light of this lapse, would ICAO have to think about a separate Annex for both types of cabin crew? It must be mentioned that the lack of attention paid by the aviation community to the importance of the flight attendant's role in a commercial flight has led to recurring instances of breakdown of communication between cabin crew and technical crew. Inevitably, this anomaly may pose serious problems in the area of air carrier liability.

Aircraft operations are addressed in Annex 6 to the Convention and is a vital regulatory aspect of safe aircraft operations. Prudent operation of aircraft is vital in avoiding accidents and incidents. Annex 6 addresses aeronautical aspects of the operations of aircraft. The essence of Annex 6, simply put, is that the operation of aircraft engaged in international air transport must be as standardized as possible to ensure the highest levels of safety and efficiency. Incontrovertibly, this would form a key focus in commercial space transport. In all phases of aircraft operations, minimum standards are the most acceptable compromise as they make commercial and general aviation viable without prejudicing safety. The Standards accepted by all Contracting States cover such areas as aircraft operations, performance, communications and navigation equipment, maintenance, flight documents, responsibilities of flight personnel and the security of the aircraft. The advent of the turbine engine and associated high performance aircraft designs necessitated a new approach to civil aircraft operation. Aircraft performance criteria, flight instruments, navigation equipment and many other operational aspects required new techniques, and they in turn created the need for international regulations to provide for safety and efficiency.

Annex 12 to the Chicago Convention requires Contracting States to coordinate their search and rescue (SAR) organizations with those of neighbouring Contracting States[9] with a recommendation that such States should, whenever necessary, coordinate their SAR operations with those of neighbouring States[10] and develop common SAR procedures to facilitate coordination of SAR operations with those of neighbouring States.[11] These provisions collectively call upon all Contracting States to bond together in coordinating both their SAR organizations and operations. In the air transport field, the dilemma facing many States extending both to airports and airlines, relates to the lack of rapid response, adequate equipment and well-trained crews, all of which are critical to passenger survival in the event of an aircraft disaster. Would these principles be adapted to cover spaceports? As a later discussion will address, there are corresponding provisions in the space law regime as well, regarding SAR. Although most States are particularly mindful of these compelling needs, they are by no means confined to the a particular region. An example of this crisis can be cited with the 1980 incident of a Saudi Arabian Airlines L-1011 catching fire shortly after leaving Riyadh Airport. Although the pilot turned back for an emergency landing and made a perfect touchdown, nearly 30 min passed before firemen managed to go in, by which time all passengers and crew had perished. This could have been a survivable accident.[12] To the contrary, a hijacking incident involving a Boeing 767 aircraft on the shores of Comoros, in November 1996, when the aircraft crashed due to lack of fuel, showed how spontaneous reaction from even non-trained professionals at rescue efforts could help. In this instance, the quick response of tourists at the scene ensured that 51 of the 175 passengers on board were saved.[13]

Another example would be accident investigation in space flight. Annex 13 to the Chicago Convention covers accident investigations regarding aircraft. The Annex provides the international requirements for the investigation of aircraft

---

[9] Standard 3.1.1.

[10] Recommendation 3.1.2.

[11] Recommendation 3.1.2.1.

[12] David Morrow, Preparing for Disaster, Airport Support, April 1995 at p. 29.

[13] Report in FAZ No. 275/1996 (25 November 1996) at p. 9.

accidents and incidents. It has been written in a way that can be understood by all participants in an investigation. As such, it serves as a reference document for people around the world who may be called on, often without any lead time, to deal with the many aspects involved in the investigation of an aircraft accident or serious incident. As an example, the Annex spells out which States may participate in an investigation, such as the States of Occurrence, Registry, Operator, Design and Manufacture. It also defines the rights and responsibilities of such States.

Fatigue risk management and its regulation would form an integral part of safety in commercial air transport. Its adaptability from corresponding regulatory standards in the air transport field would be a key consideration. Appendix 8 to Annex 6 Part 1, gives detailed requirements that the operator is expected to comply with. At the outset, the Appendix requires the operator to define its FRMS policy, with all elements of the FRMS clearly identified.[14] The next provision—para 1.1.2—provides that the operator's FRMS policy require that the scope of the FRMS operations be clearly defined in the Operations Manual. Paragraph 1.1.3. prescribes various elements to be included in the policy where the policy should: reflect the shared responsibility of management, flight and cabin crews, and other involved personnel; clearly state the safety objectives of the FRMS; be signed by the accountable executive of the organization; be communicated, with visible endorsement, to all the relevant areas and levels of the organization; declare management commitment to effective safety reporting; declare management commitment to the provision of adequate resources for the FRMS; declare management commitment to continuous improvement of the FRMS; require that clear lines of accountability for management, flight and cabin crews, and all other involved personnel are identified; and require periodic reviews to ensure it remains relevant and appropriate.

The Appendix also requires an operator to develop and keep current FRMS documentation that describes and records: FRMS policy and objectives; FRMS processes and procedures; accountabilities, responsibilities and authorities for these processes and procedures; mechanisms for ongoing involvement of management, flight and cabin crew members, and all other involved personnel; FRMS training programs, training requirements and attendance records; scheduled and actual flight times, duty periods and rest periods with significant deviations and reasons for deviations noted; and FRMS outputs including findings from collected data, recommendations, and actions taken.

Space crew integrity would be another issue for consideration. Annex 6 (Operation of Aircraft) to the Chicago Convention provides that:

> The pilot-in-command shall be responsible for the operation and safety of the aeroplane and for the safety of all persons on board, during flight time.[15]

---

[14] Paragraph 1.1.1.

[15] See Annex 6 to the Convention on International Civil Aviation signed in Chicago on 7 December 1944, para 4.5.1.

This presumption of responsibility has influenced some States which have signed or ratified the Convention and is reflected clearly in their air navigation laws.[16] These laws have been have been observed to list requirements which any pilot with a sense of good airmanship would naturally comply with. Failure to comply with such regulations has been clearly interpreted to be bad airmanship which renders the pilot liable for prosecution on a criminal charge.[17] In any event, the fundamental postulate which imposes *prima facie* responsibility on the pilot has been accepted as a general principle of liability of the pilot which sets the base for determining his legal status and responsibility.[18]

The legal responsibility placed on the commander of the aircraft is therefore inextricably linked with the expectation of good airmanship. Airmanship has been regarded as an indefinable quality and has been used to describe the intuitive faculty of the pilot where he concerns himself with what is right or wrong in the operation of an aircraft which is acquired by sustained experience in flying.[19] Needless to say, a pre-existing medical condition such as depression could adversely affect the judgment of a pilot and preclude him from exercising good airmanship.

4.3.2 Spaceports

On to the subject of another important consideration in commercial space transport —spaceports. The subject of aerodromes is addressed in Annex 14 to the Chicago Convention. A distinction of Annex 14 is the broad range of subjects it contains. It extends from the planning of airports and heliports to such details as switch-over times for secondary power supply; from civil engineering to illumination engineering; from provision of sophisticated rescue and fire fighting equipment to simple requirements for keeping airports clear of birds. The impact of these numerous subjects on the Annex is compounded by the rapidly changing industry

---

[16] *See The British Air Navigation* Order (1985) Article 32, which states, *inter alia:*
  The Commander of an aircraft registered in the United Kingdom shall satisfy himself before the aircraft takes off—that the flight can safely be made, taking into account the latest information available as to the route and aerodromes to be used, the weather reports and forecasts available and any alternative course of action which can be adopted in case the flight cannot be completed as planned; See also generally, *U.S. Federal Aviation Regulations* FAR 91.3(a), Australian Air Navigation Regulations, Regulation 219 and New Zealand Civil Aviation Regulations (1953), Regulation 59.

[17] See N. Price, *Pilot Error* (1976) at pp. 238–239. See also generally the findings of the *New Zealand Royal Commission of Inquiry into the 1979 Aft. Erebus DCIO Disaster.*

[18] Abeyratne 1998, pp. 219–231. See also N.M. Matte, *The International Legal Status of Aircraft Commander* (1975) at 34 and VidelaEscalada, *Aeronautical Law* (1979) at 2lO-211, S.M. Speiser. & C.F. Krause, *Aviation Tort Law (Vol. 1, 1978)* 473.

[19] A.J. Burridge, The Dismissal of a Pilot for Poor airmanship—The Employer's Point of View, *Aeronautical Journal* May 1977, 206.

which airports must support. New aircraft models, increased aircraft operations, operations in lower visibilities and technological advances in airport equipment combine to make Annex 14 one of the most rapidly changing Annexes.

In addition to the obligation of the State to provide certain services as enumerated in Article 28 of the Chicago Convention, responsibility of the State would also extend to the provision of accurate air traffic control services at the aerodrome. States have to be mindful of the fact that their overall responsibility under the Chicago Convention in providing air navigation services extends to the air traffic controller, whose service is of a unique nature. The special feature in the provision of air traffic control is brought to bear by the nature of the service provided, be it in the relaying of information on meteorology or on traffic. Globally, air traffic control services offer information relayed by people by means of radio communication involving extremely short time periods and using a standard set of terminology in the English language, even in regions of the world where English is not the first language.[20]

The provision of meteorological information to airports and aircraft about to land or take off is also part of State responsibility. Annex 3 to the Chicago Convention provides in Standard 2.1.1. that the objective of meteorological service for international air navigation shall be to contribute towards the safety, regularity and efficiency of international air navigation. This objective shall be achieved by supplying the following users: operators, flight crew members, air traffic services units, search and rescue services units, airport managements and others concerned with the conduct or development of international air navigation, with the meteorological information necessary for the performance of their respective functions.[21]

State responsibility for the provision of meteorological information is provided for in Standard 2.1.4. where each Contracting State is required to ensure that the designated meteorological authority complies with the requirements of the World Meteorological Organization in respect of qualifications and training of meteorological personnel providing service for international air navigation.[22]

It is also provided in the Annex that close liaison shall be maintained between those concerned with the supply and those concerned with the use of meteorological information on matters which affect the provision of meteorological service for international air navigation.[23] Furthermore, States have responsibility establish one or more aerodrome and/or other meteorological offices which shall be adequate for the provision of the meteorological service required to satisfy the needs of international air navigation.[24]

---

[20] Miyagi 2005, p. 143.

[21] Standard 2.1.2.

[22] Requirements concerning qualifications and training of meteorological personnel in aeronautical meteorology are given in WMO Publication No. 49, Technical Regulations, Volume I— General Meteorological Standards and Recommended Practices, Chapter B.4—*Education and Training*.

[23] Standard 2.2.1.

[24] Standard 3.3.1.

It is incontrovertible that the responsibility of the State is not extinguished merely because an airport is subject to private ownership or private management control. In international air transport, the mere fact that the State has to provide airport services under Article 28 of the Chicago Convention and indeed designate airports within its territory for landing purposes as per Articles 10 and 68 thereof imposes legal responsibility upon the State to be accountable at public international law for any liability incurred as a result of action on the part of airports within its territory.

Irrespective of the responsibility of a State with regard to airports within its territories, which is founded both at customary international law and at private law for liability incurred by airports, a privately run airport may incur tortuous liability on a private basis, as the occupier of the premises. In the instance of a privately managed airport where the entity charged with managing airport services is located within the airport premises, such an entity would be considered as a legal occupier for purposes of liability.

4.3.3 Cargo

It is well known that both at present, and in the future, the carriage of cargo and equipment on board spacecraft is a regular activity. Annex 18 (The Safe Transport of Dangerous Goods by Air) to the Chicago Convention specifies the broad Standards and Recommended Practices to be followed to enable dangerous goods to be carried safely. The Annex contains fairly stable material requiring only infrequent amendment using the normal Annex amendment process. The Annex also makes binding upon Contracting States the provisions of the Technical Instructions, which contain the very detailed and numerous instructions necessary for the correct handling of dangerous cargo. These require frequent updating as developments occur in the chemical, manufacturing and packaging industries, and a special procedure has been established by the Council to allow the Technical Instructions to be revised and reissued regularly to keep up with new products and advances in technology.

The ICAO requirements for dangerous goods have been largely developed by a panel of experts which was established in 1976. This panel continues to meet and recommends the necessary revisions to the Technical Instructions. As far as possible the Technical Instructions are kept aligned with the recommendations of the United Nations Committee of Experts on the Transport of Dangerous Goods and with the regulations of the International Atomic Energy Agency. The use of these common bases by all forms of transport allows cargo to be transferred safely and smoothly between air, sea, rail and road modes.

The ICAO requirements for the safe handling of dangerous goods firstly identify a limited list of those substances which are unsafe to carry in any circumstances and then show how other potentially dangerous articles or substances can be transported safely. The nine hazard classes are those determined by the United Nations Committee of Experts and are used for all modes of transport.

Class 1 includes explosives of all kinds, such as sporting ammunition, fireworks and signal flares. Class 2 comprises compressed or liquefied gases which may also be toxic or flammable; examples are cylinders of oxygen and refrigerated liquid nitrogen. Class 3 substances are flammable liquids including gasoline, lacquers, paint thinners, etc. Class 4 covers flammable solids, spontaneously combustible materials and materials which, when in contact with water, exit flammable gases (examples are some powdered metals, cellulose type film and charcoal).

Class 5 covers oxidizing material, including bromates, chlorates or nitrates; this class also covers organic peroxides which are both oxygen carriers and very combustible. Poisonous or toxic substances, such as pesticides, mercury compounds, etc., comprise Class 6, together with infectious substances which must sometimes be shipped for diagnostic or preventative purposes. Radioactive materials are in Class 7; these are mainly radioactive isotopes needed for medical or research purposes but are sometimes contained in manufactured articles such as heart pacemakers or smoke detectors. Corrosive substances which may be dangerous to human tissue or which pose a hazard to the structure of an aircraft are dealt with in Class 8 (for example, caustic soda, battery fluid, paint remover). Finally, Class 9 is a miscellaneous category for other materials which are potentially hazardous in air transport, such as magnetized materials which could affect the aircraft's navigational systems.

The security of cargo carried would also be a critical consideration in commercial space transportation and ICAO material could be considered relevant in formulating guidelines. In this regard Annex 17 contains extracts from Annex 18 which require each Contracting State to take necessary measures to achieve compliance with the detailed provisions contained in the Technical Instructions for the Safe Transport of Dangerous Goods by Air (Doc 9284), which are approved and issued periodically in accordance with procedure established by the ICAO Council. The requirement also covers compliance with any amendment to the Technical Instructions.[25]

Annex 17 to the Chicago Convention also contains some general provisions that may apply to the illegal carriage by air of infectious pathogens. Standard 5.1.2 devolves responsibility upon Contracting States to ensure that, when reliable information exists that an aircraft may be subject to an act of unlawful interference, that the aircraft is searched for illegal weapons, explosives and other dangerous devices. The main preventive objective contained in Standard 4.1 which ensures that States establish measures to prevent weapons, explosives or any other dangerous devices articles or substances which may be used to commit an act of unlawful interference and which are not authorized, from being carried on board.

---

[25] See Attachment to Annex 17 to the Chicago Convention, at ATT-11, which reproduces Standard 2.2.1 and Recommendation 2.2.2 of Annex 18. Recommendation 2.2.2 provides that each Contracting State should inform ICAO of difficulties encountered in the application of the Technical Instructions and of any amendments which it would be desirable to make to them.

Annex 18—on the safe transport of dangerous goods by air—applies to all international operations of civil aircraft and forbids, in Standard 4.1, the transport of dangerous goods by air except as established in the Annex and detailed specifications and procedures provided in the Technical Instructions. he Annex was developed by the Air Navigation Commission of the Organization in response to a need expressed by States for an internationally agreed set of provisions governing the safe transport of dangerous goods by air. The Annex draws the attention of the States to the need to adhere to Technical Instructions for the Safe Transport of Dangerous Goods by Air[26] developed by ICAO, according to which packaging used for the transportation of dangerous goods by air shall be of good quality and shall be constructed and securely closed so as to prevent leakage[27] and labelled with the appropriate labels.[28]

Annex 18 clearly identifies in Chapter 8 requirements that the carrier has to comply with when accepting dangerous goods for transport. According to these requirements the operator has to ensure that dangerous goods are accompanied by a completed dangerous goods transport document, except when the Technical Instructions indicate that such a document is not required.[29] The carrier is also required not to accept dangerous goods until the package, over pack or freight container containing the dangerous goods has been inspected in accordance with acceptance procedures contained in the Technical Instructions.[30]

More specifically, the Annex has specific provisions concerning acceptance of radioactive materials, according to which there is a requirement presumably to be complied with by both the customs authorities and the carrier that packages and over packs containing dangerous goods and freight containers containing radioactive materials shall not be loaded into a unit load device or an aircraft for carriage before they have been inspected for evidence of leakage or damage[31] It goes on to say that a unit load device shall not be loaded aboard an aircraft unless the device has been inspected and found free from any evidence of leakage from, or damage to, any dangerous goods contained therein.[32]

The Instructions are a critical contribution of ICAO to the subject of dangerous goods and safety in air transport. The provisions contained therein prescribe the detailed requirements applicable to the international civil transport of dangerous

---

[26] Technical Instructions for the Safe Transport of Dangerous Goods by Air, Doc 9284 AN/905 2007–2008 Edition (hereafter referred to as *the Instructions*). The Technical Instructions are quite specific and comprehensive. For a detailed discussion of *the Instructions* see Warner and Rooney 1997, pp. 23–24 and 29 at 23.

[27] Annex 18 to the Convention on International Civil Aviation (The Safe Transport of Dangerous Goods by Air), Second Edition—July 1989, Standard 5.2.1.

[28] *Id.* Standard 6.1.

[29] *Id.* Standard 8.1(a).

[30] *Id.* standard 8.1(b).

[31] *Id.* Standard 8.4.1.

[32] *Id.* Standard 8.4.2.

goods by air.[33] The overarching principle of *the Instructions* is that any substance which, as presented for transport, is liable to explode, dangerously react, produce a flame or dangerous emission of heat or toxic, corrosive or flammable gases or vapours under conditions normally encountered in transport must not be carried in aircraft under any circumstances.[34]

4.3.4 Environment

One of the key issues for deliberation by the ICAO Council in the context of commercial space transport would be spaceports and surrounding noise. The balancing of airport development and ecological considerations i.e. city planning, noise pollution avoidance, is very much a part of ICAO's regulatory role in issues related to the effects of international civil aviation on the environment. The ICAO *Airport Planning Manual*[35] ensures a balance between airport development and ecological considerations and includes findings of ICAO on aviation and the environment.

In its findings, ICAO records that studies of air quality at certain large airports and nearby areas reflect the fact that automobiles, airport ground vehicles and other urban pollution sources account for most of the atmospheric pollution[36] and that airports may destroy the natural habitat and feeding grounds of wild life and may eradicate or deplete certain flora important to the ecological balance of the area.[37] The Manual also considers the necessity to avoid contamination of rivers and streams by airport waste disposal and drainage systems,[38] the avoidance of noise caused by aircraft to human habitation[39] and highway planning.[40] Finally, the document calls for a detailed study of the impact of airport development on the environment in the form of an environmental impact statement.[41]

In terms of noise, ICAO has been conscious of noise levels of sub sonic and supersonic aircraft in the vicinity of airports—a debate that might be renewed in the context of spaceports. At the 28th Assembly Sessions held in October 1990, the ICAO Assembly observed that while certification standards for subsonic jet aircraft noise levels are specified in Volume 1, Chapter 2 and Chapter 3 of Annex 16 and

---

[33] *Id.* 1.1.1. at p. 1-1-1.

[34] *Id.* 2.1 at p. 1-2-1. This excludes items such as aerosols, alcoholic beverages, perfumes, colognes safety matches and liquefied gas lighters carried on board by the operator for use or sale.

[35] See *Airport Planning Manual* Doc 9184—AN 902 Part 2, 13.2.

[36] *Id.* 2.1.3.

[37] *Id.* 2.2.1.

[38] *Id.* 2.4.1.

[39] *Id.* 2.5.2.

[40] *Id.* 4.3.1.

[41] *Id.* 2.6.1.

that environmental problems due to aircraft noise continued to exist in the neigh-bourhood of many international airports, some States were consequently consid-ering restrictions on the operations of aircraft which exceed the noise levels in Volume I, Chapter 3 of Annex 16. The Assembly also recognized that the noise standards in Annex 16 were not intended to introduce operating restrictions on aircraft and that operating restrictions on existing aircraft would increase the costs of airlines and would impose a heavy economic burden, particularly on those airlines which do not have the financial resources to re-equip their fleets. Therefore, considering that resolution of problems due to aircraft noise must be based on the mutual recognition of the difficulties encountered by States and a balance among their different concerns, the Assembly, by Resolution A 28-3, urged States not to introduce any new operating restrictions on aircraft which exceed the noise levels in Volume I, Chapter 3 of Annex 16 before considering certain key issues.

The Assembly, while urging States, if and when any new noise certification standards are introduced which are more stringent than those in Volume I, Chapter 3 of Annex 16, not to impose any operating restrictions on Chapter 3 compliant aircraft, urged the Council to promote and States to develop an integrated approach to the problem of aircraft noise, including land-use planning procedures around international airports, so that any residential, industrial or other land-use that might be adversely affected by aircraft noise is minimal. The Assembly further urged States to assist aircraft operators in their efforts to accelerate fleet modernization and thereby prevent obstacles and permit all States to have access to lease or purchase aircraft compliant with Chapter 3, including the provision of multilateral technical assistance where appropriate.

With regard to emissions of spacecraft, this might be a consideration for the future in terms of regulation. One commentator opines:

> Although the impacts are far fewer than those caused by the aviation industry, each object sent to space has an impact on Earth's air quality and atmosphere. Launch activities and the propellants released can lead to environmental degradation. However, launching activities and rocket emissions generally are not included in environmental assessments; and new rocket propulsion systems, such as hybrid propellants and hypersonic propulsion, are being developed and promoted without due regard to their possible environmental impacts. The reason is that the contribution of space activities in atmospheric pollutions at the current rate does not alarm environmentalists so that they would take serious action.[42]

ICAO would be well advised to remand the issue of spacecraft emissions to its panel of experts in the Council's Committee on Aviation Environmental Protection (CAEP) when the need arises for a technical assessment.

---

[42] Seyedeh Mahboubeh Mousavi Sameh, Suborbital Flights: Selected Legal Issues, A thesis submitted to McGill University (2013), 68. The author also states: "Suborbital vehicle emissions and noise problems will basically depend on the engines and the kinds of designs deployed. Even if alleged to be environmental friendly, the engines designed for these vehicles cannot be totally emission free. All propellant types and different kinds of fuels contribute to environmental pol-lution. Even water vapour emissions have ozone depletion and climate change effects. In other words, no engine type is absolutely environmentally friendly" *Id.* 71.

4.3.5 Responsibility of the Launching State for Private Acts of Individuals in Outer Space

One of the subjects that would be pertinent to the debate in expanding the scope of ICAO to commercial activities in outer space is the responsibility of the launching State (the State responsible for launching its spacecraft into outer space) for private acts of Individuals. Here, prevailing current international law principles would be relevant and persuasive. The State concerned will have to demonstrate that either it did not tolerate the offence or that it ensured the punishment of the offender. One view is that proof of such breach would lie in the causal connection between the private offender and the State. In this context, the act or omission on the part of a State is a critical determinant particularly if there is no specific intent. Generally, it is not the intent of the offender that is the determinant but the failure of a State to perform its legal duty in either preventing the offence (if such was within the purview of the State) or in taking necessary action with regard to punitive action or redress.

Finally, there are a few principles that have to be taken into account when determining State responsibility for private acts of individuals. Firstly, there has to be either intent on the part of the State towards complicit or negligence reflected by act or omission. Secondly, where condonation is concerned, there has to be evidence of inaction on the part of the State in prosecuting the offender. Thirdly, since the State as an abstract entity cannot perform an act in itself, the imputability or attribution of State responsibility for acts of its agents has to be established through a causal nexus that points the finger at the State as being responsible. For example, The International Law Commission, in Article 4 of its Articles of State Responsibility states that the conduct of any State organ which exercises judicial, legislative or executive functions could be considered an act of State and as such the acts of such organ or instrumentality can be construed as being imputable to the State. This principle was endorsed in 1999 by the ICJ which said that according to well established principles of international law, the conduct of any organ of a state must be regarded as an act of State.

The idea that States have a responsibility to ensure that victims of crime and other acts are compensated is not confined to Europe. The US Department of Justice has long had an Office for Victims of Crime [OVC] which oversees the schemes in individual States and in collaboration with the State Department, has compiled and updated a Directory of schemes in 35 countries principally for the information of US citizens who travel or reside overseas.

The law of State responsibility for private acts of individuals has evolved through the years, from being a straightforward determination of liability of the State and its agents to a rapidly widening gap between the State and non State parties. In today's world private entities and persons could wield power similar to that of a State, bringing to bear the compelling significance and modern relevance of the agency nexus between the State and such parties. This must indeed make States more aware of their own susceptibility.

The fundamental issue in the context of State responsibility for the purposes of this article is to consider whether a State should be considered responsible for its own failure or non-feasance to prevent a private act or whether the conduct of the State itself can be impugned by identifying a nexus between the perpetrator's conduct and the State. One view is that an agency paradigm, which may in some circumstances impute to a state reprehensibility on the ground that a principal-agent relationship between the State and the perpetrator existed, can obfuscate the issue and preclude one from conducting a meaningful legal study of the State's conduct.[43]

4.3.5.1 The Theory of Complicity

At the core of the principal-agent dilemma is the theory of complicity, which attributes liability to a State that was complicit in a private act. Hugo Grotius (1583–1645), founder of the modern natural law theory, first formulated this theory based on State responsibility that was not absolute. Grotius' theory was that although a State did not have absolute responsibility for a private offence, it could be considered complicit through the notion of *patienta* or *receptus*.[44] While the concept of *patienta* refers to a State's inability to prevent a wrongdoing, *receptus* pertains to the refusal to punish the offender.

The 18th Century philosopher Emerich de Vattel was of similar view as Grotius, holding that responsibility could only be attributed to the State if a sovereign refuses to repair the evil done by its subjects or punish an offender or deliver him to justice whether by subjecting him to local justice or by extraditing him.[45] This view was to be followed and extended by the British jurist Blackstone a few years later who went on to say that a sovereign who failed to punish an offender could be considered as abetting the offence or of being an accomplice.[46]

A different view was put forward in an instance of adjudication involving a seminal instance where the Theory of Complicity and the responsibility of states for private acts of violence was tested in 1925. The case[47] involved the Mexico-United States General Claims Commission which considered the claim of the United States on behalf of the family of a United States national who was killed in a Mexican mining company where the deceased was working. The United States argued that the Mexican authorities had failed to exercise due care and diligence in apprehending and prosecuting the offender. The decision handed down by the Commission distinguished between complicity and the responsibility to punish and

---

[43] Caron 1998, pp. 153–154 cited in Tal Becker, Terrorism and the State, Hart Monographs in Transnational and International Law, Hart Publishing: 2006, at p. 155.

[44] H Grotius, JB Scott, (tr), 2 *De Jure Belli Ac Pacis* (1646), pp. 523–526.

[45] De Vattel and Fenwick 1916, p. 72.

[46] Blackstone and Morrison 2001, p. 68.

[47] *Laura M.B. Janes (USA)* v. *United Mexican States* (1925) 4 R Intl Arb Awards 82.

the Commission was of the view that Mexico could not be considered an accomplice in this case.

The Complicity Theory, particularly from a Vattellian and Blackstonian point of view is merely assumptive unless put to the test through a judicial process of extradition. In this Context it becomes relevant to address the issue through a discussion of the remedy.

4.3.5.2 The Condonation Theory

The emergence of the Condonation Theory was almost concurrent with the *Jane* case[48] decided in 1925 which emerged through the opinions of scholars who belonged to a school of thought that believed that States became responsible for private acts of violence not through complicity as such but more so because their refusal or failure to bring offenders to justice was tantamount to ratification of the acts in question or their condonation.[49] The theory was based on the fact that it is not illogical or arbitrary to suggest that a State must be held liable for its failure to take appropriate steps to punish persons who cause injury or harm to others for the reason that such States can be considered guilty of condoning the criminal acts and therefore become responsible for them.[50] Another reason attributed by scholars in support of the theory is that during that time, arbitral tribunals were ordering States to award pecuniary damages to claimants harmed by private offenders, on the basis that the States were being considered responsible for the offences.[51]

The responsibility of governments in acting against offences committed by private individuals may sometimes involve condonation or ineptitude in taking effective action against terrorist acts, in particular with regard to the financing of terrorist acts. The United Nations General Assembly, on 9 December 1999, adopted the International Convention for the Suppression of the Financing of Terrorism,[52] aimed at enhancing international co-operation among States in devising and adopting effective measures for the prevention of the financing of terrorism, as well as for its suppression through the prosecution and punishment of its perpetrators.

The Convention, in its Article 2 recognizes that any person who by any means directly or indirectly, unlawfully or wilfully, provides or collects funds with the intention that they should be used or in the knowledge that they are to be used, in full or in part, in order to carry out any act which constitutes an offence under certain named treaties, commits an offence. One of the treaties cited by the

---

[48] *Id.*

[49] *Black's Law Dictionary* defines condonation as "pardon of offense, voluntary overlooking implied forgiveness by treating offender as if offense had not been committed".

[50] Jane's case, *Supra*, note 47, at p. 92.

[51] Hyde (1928), pp. 140–142.

[52] International Convention for the Suppression of the Financing of Terrorism, adopted by the General Assembly of the United Nations in resolution 54/109 of 9 December 1999.

Convention is the International Convention for the Suppression of Terrorist Bombings, adopted by the General Assembly of the United Nations on 15 December 1997.[53]

The Convention for the Suppression of the Financing of Terrorism also provides that, over and above the acts mentioned, providing or collecting funds toward any other act intended to cause death or serious bodily injury to a civilian, or to any other person not taking an active part in the hostilities in the situation of armed conflict, when the purpose of such act, by its nature or context, is to intimidate a population, or to compel a government or an international organization to do or to abstain from doing any act, would be deemed an offence under the Convention.

The United Nations has given effect to this principle in 1970 when it proclaimed that:

> Every State has the duty to refrain from organizing or encouraging the organization of irregular forces or armed bands, including mercenaries, for incursion into the territory of another State. Every State has the duty to refrain from organizing, instigating, assisting or participating in acts of civil strife or terrorist acts in another State or acquiescing in organized activities within its territory directed towards the commission of such acts, when the acts referred to in the present paragraph involve a threat or use of force.[54]

Here, the words *encouraging* and *acquiescing in organized activities within its territory directed towards the commission of such acts* have a direct bearing on the concept of condonation and would call for a discussion about how States could overtly or covertly encourage the commission of such acts. One commentator[55] identifies three categories of such support: *Category I* support entails protection, logistics, training, intelligence, or equipment provided terrorists as a part of national policy or strategy; *Category II* support is not backing terrorism as an element of national policy but is the toleration of it; *Category III* support provides some terrorists a hospitable environment, growing from the presence of legal protections on privacy and freedom of movement, limits on internal surveillance and security organizations, well-developed infrastructure, and émigré communities

Another commentator[56] discusses what he calls the *separate delict theory*' in State responsibility, whereby the only direct responsibility of the State is when it is responsible for its own wrongful conduct in the context of private acts, and not for the private acts themselves. He also contends that indirect State responsibility is occasioned by the State's own wrongdoing in reference to the private terrorist conduct. The State is not held responsible for the act of terrorism itself, but rather for its failure to prevent and/or punish such acts, or for its active support for or

---

[53] A/52/653, 25 November 1997.

[54] Declaration on Principles of International Law Concerning Friendly Relations and Co-operation Among States in Accordance with the Charter of the United Nations, UN General Assembly Resolution 2625 (XXV) 24 October 1970.

[55] Steven Metz, State Support for Terrorism, Defeating Terrorism, Strategic Issue Analysis, at http://www.911investigations.net/IMG/pdf/doc-140.pdf.

[56] Becker 2006.

acquiescence in terrorism.[57] Arguably the most provocative and plausible feature in this approach is the introduction by the commentator of the desirability of determining State liability on the theory of causation. He emphasizes that:

> The principal benefit of the causality based approach is that it avoids the automatic rejection of direct State responsibility merely because of the absence of an agency relationship. As a result, it potentially exposes the wrongdoing State to a greater range and intensity of remedies, as well as a higher degree of international attention and opprobrium for its contribution to the private terrorist activity.[58]

The causality principle is tied in with the rules of State Responsibility enunciated by the International Law Commission and Article 51 of the United Nations Charter which states that nothing in the Charter will impair the inherent right of individual or collective self-defense if an armed attack occurs against a Member of the United Nations, until the Security Council has taken measures necessary to maintain international peace and security. The provision goes on to say that measures taken by Members in the exercise of this right of self-defense will be immediately reported to the Security Council and will not in any way affect the authority and responsibility of the Security Council under the present Charter to take at any time such action as it deems necessary in order to maintain or restore international peace and security.

The International Law Commission has established that a crime against the peace and security of mankind entails individual responsibility, and is a crime of aggression.[59] A further link lies in the Rome Statute of the International Criminal court, which defines a war crime, *inter alia*, as intentionally directing attacks against civilian objects; attacking or bombarding, by whatever means, towns, villages, dwellings or buildings which are undefended and which are not military objects; employing weapons, projectiles, and Materials and methods of warfare that cause injury.[60] The Statute also defines as a war crime, any act which is intentionally directed at buildings, material, medical units and transport, and personnel.

4.3.5.3 The Role of Knowledge

Another method of determining State responsibility lies in the determination whether a State had actual or presumed knowledge of acts of its instrumentalities, agents or private parties which could have alerted the State to take preventive action. International responsibility of a State cannot be denied merely on the strength of the claim of that State to sovereignty. Although the Chicago Convention in Article 1 stipulates that the contracting States recognize that every State has

---

[57] *Id.* Chapter 2, 67.

[58] Becker, *supra* note 171, at p. 335.

[59] Draft Code of Crimes Against the Peace and Security of Mankind, International Law Commission Report, 1996, Chapter II Article 2.

[60] Rome Statute of the International Criminal Court, Article 8.2(b)(ii), (V) and (XX).

complete and exclusive sovereignty over the airspace above its territory, the effect of this provision cannot be extended to apply to State immunity from responsibility to other States. Professor Huber in the *Island of Palmas* case[61] was of the view:

> Sovereignty in the relations between States signifies independence. Independence in regard to a portion of the globe is the right to exercise therein, to the exclusion of any other State, the functions of a State…Territorial sovereignty…involves the exclusive right to display the activities of a State.[62]

Professor Huber's definition, which is a simple statement of a State's rights, has been qualified by Starke as the residuum of power which a State possesses within the confines of international law.[63] Responsibility would devolve upon a State which launches a spacecraft to other States that are threatened by such acts. The International Court of Justice (ICJ) recognised in the *Corfu Channel* Case:

In the famous *Corfu Channel* case, the International Court of Justice applied the subjective test and applied the fault theory. The Court was of the view that:

> It cannot be concluded from the mere fact of the control exercised by a State over its territory and waters that the State necessarily knew, or ought to have known, of any unlawful act perpetrated therein, nor yet that it necessarily knew, or should have known the authors. This fact, by itself and apart from other circumstances, neither involves prima facie responsibility nor shifts the burden of proof.[64]

The Court, however, pointed out that exclusive control of its territory by a State had a bearing upon the methods of proof available to establish the involvement or knowledge of that State as to the events in question.

Apart from the direct attribution of responsibility to a State, particularly in instances where a State might be guilty of a breach of treaty provisions, or violate the territorial sovereignty of another State, there are instances where an act could be imputed to a State.[65] Imputability or attribution depends upon the link that exists between the State and the legal person or persons actually responsible for the act in question. The legal possibility of imposing liability upon a State wherever an

---

[61] The *Island of Palmas* Case (1928) 11 U.N.R. I.A.A. at p. 829.

[62] *Ibid.*

[63] Starke 1989, p. 3.

[64] The *Corfu Channel* Case, ICJ Reports, 1949, p. 4.

[65] There are some examples of imputability, for example the incident in 1955 when an Israeli civil aircraft belonging to the national carrier El Al was shot down by Bulgarian fighter planes, and the consequent acceptance of liability by the USSR for death and injury caused which resulted in the payment of compensation to the victims and their families. See 91 *ILR* 287. Another example concerns the finding of the International Court of Justice that responsibility could have been be imputed to the United States in the *Nicaragua* case, where mines were laid in Nicaraguan waters and attacks were perpetrated on Nicaraguan ports, oil installations and a naval base by persons identified as agents of the United States. See *Nicaragua* v. *the United States*, ICJ Reports 1986, 14. Also, 76 *ILR* 349. There was also the instance when the Secretary General of the United Nations mediated a settlement in which a sum, *inter alia* of $7 million was awarded to New Zealand for the violation of its sovereignty when a New Zealand vessel was destroyed by French agents in New Zealand. See the *Rainbow Warrior* case, 81 *AJIL*, 1987 at 325. Also in 74 *ILR* at p. 241.

official could be linked to that State encourages a State to be more cautious of its responsibility in controlling those responsible for carrying out tasks for which the State could be ultimately held responsible. In the same context, the responsibility of placing mines was attributed to Albania in the *Corfu Channel* case since the court attributed to Albania the responsibility, since Albania was known to have knowledge of the placement of mines although it did not know who exactly carried out the act. It is arguable that, in view of the responsibility imposed upon a State by the Chicago Convention on the provision of air navigation services, the principles of immutability in State responsibility could be applied to an instance of an act or omission of a public or private official providing air navigation services.

The sense of international responsibility that the United Nations ascribed to itself had reached a heady stage at this point, where the role of international law in international human conduct was perceived to be primary and above the authority of States. In its Report to the General Assembly, the International Law Commission recommended a draft provision which required:

> Every State has the duty to conduct its relations with other States in accordance with international law and with the principle that the sovereignty of each State is subject to the supremacy of international law.[66]

This principle, which forms a cornerstone of international conduct by States, provides the basis for strengthening international comity and regulating the conduct of States both internally—within their territories—and externally, towards other States. States are effectively precluded by this principle of pursuing their own interests untrammelled and with disregard to principles established by international law.

The above discussion leads one to conclude that the responsibility of a State for private acts of individuals which is determined by the quantum of proof available that could establish intent or negligence of the State, which in turn would establish complicity or condonation on the part of the State concerned. One way to determine complicity or condonation is to establish the extent to which the State adhered to the obligation imposed upon it by international law and whether it breached its duty to others. In order to exculpate itself, the State concerned will have to demonstrate that either it did not tolerate the offence or that it ensured the punishment of the offender. *Brownlie* is of the view that proof of such breach would lie in the causal connection between the private offender and the State.[67] In this context, the act or omission on the part of a State is a critical determinant particularly if there is no specific intent.[68] Generally, it is not the intent of the offender that is the determinant but the failure of a State to perform its legal duty in either preventing the offence

---

[66] *Report of the International Law Commission to the General Assembly on the Work of the 1st Session*, A/CN.4/13, June 9 1949, at p. 21.

[67] Brownlie 1983, p. 39.

[68] Report of the International Law Commission to the United Nations General Assembly, UNGOAR 56th Session, Supp. No. 10, *UN DOC A/56/10*, 2001 at p. 73.

(if such was within the purview of the State) or in taking necessary action with regard to punitive action or redress.[69]

Finally, there are a few principles that have to be taken into account when determining State responsibility for private acts of individuals that unlawfully interfere with activities in outer space. Firstly, there has to be either intent on the part of the State towards complicit or negligence reflected by act or omission. Secondly, where condonation is concerned, there has to be evidence of inaction on the part of the State in prosecuting the offender. Thirdly, since the State as an abstract entity cannot perform an act in itself, the imputability or attribution of State responsibility for acts of its agents has to be established through a causal nexus that points the finger at the State as being responsible. For example, The International Law Commission, in Article 4 of its Articles of State Responsibility states that the conduct of any State organ which exercises judicial, legislative or executive functions could be considered an act of State and as such the acts of such organ or instrumentality can be construed as being imputable to the State. This principle was endorsed in 1999 by the ICJ which said that according to well established principles of international law, the conduct of any organ of a state must be regarded as an act of State.[70]

The law of State responsibility for private acts of individuals has evolved through the years, from being a straightforward determination of liability of the State and its agents to a rapidly widening gap between the State and non State parties. In today's world private entities and persons could wield power similar to that of a State, bringing to bear the compelling significance and modern relevance of the agency nexus between the State and such parties. This must indeed make States more aware of their own susceptibility.

4.3.6 Responsibilities of States of Registration of Aircraft and States Launching Aircraft

By registering an aircraft a State bestows its nationality to that aircraft. Such a State takes on responsibility ascribed to it under various provisions of the Chicago Convention and its Annexes in such areas as licensing, monitoring and assurance of airworthiness by issuance of certificates, appropriate documentation carried in aircraft and jurisdiction over the high seas. In addition, general legal principles of State responsibility would apply to a State in which an aircraft is registered. The certification of pilots is also the responsibility of the State of registration.

The Chicago Convention in Article 32 requires that the pilot of every aircraft and the other members of the operating crew of every aircraft engaged in international

---

[69] de Arechaga 1968, p. 531 at p. 535.

[70] *Differences Relating to Immunity from Legal Process of a Special Rapporteur*, ICJ Reports 1999, 62 at 87.

navigation be provided with certificates of competency and licences issued or rendered valid by the State in which the aircraft is registered. ICAO's global medical standards for the issuance of a pilot's license are contained in Chapter 6 of Annex 1[71] to the Chicago Convention. The Annex provides that if the medical standards prescribed in Chapter 6 for a particular license are not met, the appropriate medical assessment shall not be issued or renewed unless there were special circumstances that led to the applicant's failure to meet such requirements and that the special abilities, skill and experience of the applicant are given due consideration and that the license is appropriately endorsed with any special limitation when the safe performance of the license holder's duties is dependent upon that limitation.[72] The Annex further goes on to provide that license holders shall not exercise the privilege of their licenses and related ratings at any time when they are aware of any decrease in their medical fitness which might render them unable to safely and properly exercise their privileges.[73] A recommendation follows, that license holders should inform the licensing authority of confirmed pregnancy or any decrease in medical fitness of a duration of more than 20 days or which requires continued treatment with prescribed medication or which has required hospital treatment.[74] Another relevant provision prescribes that license holders shall not exercise the privilege of their licenses and related ratings at any time when they are under the influence of any psychoactive substance which might render them unable to safely and properly exercise their privileges.[75] Detailed guidance for the implementation of Annex 1 is contained in the *ICAO Manual of Civil Aviation Medicine*.[76]

International regulations adopted under the auspices of ICAO require that a pilot has to have a certificate of competence issued by the State in which the aircraft he flies is registered, if he were to undertake flying an aircraft. Medical certification is an essential component in the licensing process and conditions and guidelines for the issuance of such certificates are provided in detail in ICAO documents. The overall responsibility of the pilot for the safety of his flight and that of persons therein which is legally recognized by international treaty, has necessitated the grounding of pilots for many reasons where their health did not reach the standards required, which in turn has resulted often in the concealment by pilots during their medical examinations of pre-existing illnesses.

Article 33 of the Chicago Convention provides that Contracting States will consider as valid certificates of airworthiness and certificates of competency and licenses issued or rendered valid by another contracting State in which the aircraft is registered, provided that the requirements under which such certificates or licences were issued or rendered valid are equal to or above the minimum standards which

---

[71] Personnel Licensing, 10th Edition, July 2006.

[72] *Id*, Standard 1.2.4.8.

[73] *Id*. Standard 1.2.6.1.

[74] *Id*. Recommendation 1.2.6.1.1.

[75] *Id*. Standard 1.2.7.1.

[76] *Doc 8984-AN/895*.

may be established from time to time pursuant to the Convention. The ICAO Assembly at its Second Session convened in Geneva from 1 to 21 June, 1948 adopted Resolution A2-44 (Recognition of certificates of airworthiness and certificates of competency or licenses of aircrew issued by the State of Registry of the aircraft) resolved that the Council should study Article 33 further to determine whether the Article should include elements conveying the notion that certificates of airworthiness and certificates of competency or licenses of the crew of an aircraft issued or rendered valid by a ICAO member State in which the aircraft is registered shall be considered valid by all other States for the purpose of flight of such aircraft in or across the territory of such other States, provided that after the coming into force of Standards and Recommended Practices (SARPs, i.e. after Annex 1 was adopted and came into effect) such SARPs should apply in the issuance of licenses to airmen. This was recognized subsequently at the Assembly's 21st Session through Resolution A21-21.

As for spacecraft launched by a State, its nationality, Bin Cheng states:

> Nationality would denote that the national State enjoys quasi-territorial jurisdiction as well as quasi judicial jurisfaction over any space object bearing its nationality, including all persons and things on board or constructively on board.[77]

The jurisaction is preeminent and overriding except in cases of jurisfaction pertaining to States in whose territory a spacecraft might land. A launching State would also be responsible for all activities conducted by the spacecraft in outer space. In this context a distinct link could be drawn between Article 83 bis of the Chicago Convention which admits of transfer of registration rights to a State in which an aircraft might regularly operate from, where similar alternate arrangements might be made in commercial activities of spacecraft in outer space.

4.3.7 Mining Asteroids

Another aspect of commercial space transportation emerging in recent times is mining asteroids. In April 2012, it was made known to the international community that a private entity in the United States had initiated a commercial measure that would mine asteroids. This measure, it is envisioned, would add trillions of dollars to the global GDP and open opportunities to mine precious metals such as platinum and cobalt from asteroids that are nudged to near Earth orbit. This article examines the policies of major space faring countries and international cooperation in space exploration through various studies conducted on the subject and the legal principles involved. Broadly, the theme revolves round two key issues: that space exploration must continue for technology to progress; and security issues relating to this potentially lucrative business must be looked into. Mining asteroids is a

---

[77] Cheng 1997, p. 490.

promising element in this equation that would pave the way for advances in physical and medical science as well as chemistry. Secondly, it looks at relations between the United States and China, given the fact that space diplomacy is an incipient but rapidly evolving process, and that the key to international cooperation would lie in relations between the United States and China.

In 2004, the United States, through the National Aeronautics and Space Administration(NASA) released its Vision for Space Exploration. The Vision moves towards its fundamental goal—which is to advance U.S. scientific, security, and economic interests through a robust space exploration program. To achieve this goal, the United States intends to: implement a sustained and affordable human and robotic program to explore the solar system and beyond; extend human presence across the solar system, starting with a human return to the Moon by the year 2020, in preparation for human exploration of Mars and other destinations; develop the innovative technologies, knowledge, and infrastructures both to explore and to support decisions about the destinations for human exploration; and promote international and commercial participation in exploration to further U.S. scientific, security and economic interests. The vision prompted NASA to engage, between 2004 and 2007, other space agencies in informal discussions on modalities, goals, possibilities, competencies and timeline for space exploration in the future.

This vision is not unique to the United States. The European Space Agency has its Aurora space exploration programme. China, India, Japan and Russia have ambitious national projects to explore the Moon or Mars, while future national missions are being discussed in Canada, Germany, Italy, Republic of Korea and the United Kingdom.

In 2009, the United States Human Spaceflight Plans Committee (more popularly called the Augustine Committee, named after Norman R. Augustine, Chairman) in its report[78] recognized that space exploration has become a global enterprise and that in the face of a burgeoning commercial space industry which could be encouraged to engage in space exploration, costs incurred by the government could be vastly reduced in the implementation of its space programme.[79] The Committee also opined that the United States could lead a bold new international effort in the human exploration of space with the involvement of international partners.[80]

The current vision of the leadership in the United States on space exploration, as articulated by President Obama on 15 April 2010, is that eventually there would be a manned mission on Mars.[81] President Obama has not given a time line for this occurrence and it could well be after his tenure of office, even if he were to win a

---

[78] Seeking a Human Spaceflight Program—Review of the U.S. Human Spaceflight Plans Committee, Washington D.C: October 2009.

[79] *Id.* at p. 9.

[80] The Committee was of the view that the overall NASA budget has consistently been inadequate to meet its goals and that its conclusion was that the United States was fiscally incapable of meeting its own exploration goals. See Seeking a Human Spaceflight Program Worthy of a Great Nation, *Review of U.S. Human Spaceflight Plans Committee*, October 2009 at p. 9.

[81] http://www.informationweek.com/news/storage/fabrics/224400444.

second term. This is in contrast to the declaration of President Kennedy in 1961 when he said about the moon missions:

> I believe that this nation should commit itself to achieving the goal, before this decade is out, of landing a man on the moon and returning him safely to Earth.[82]

Space exploration started as a race between the then Soviets (USSR) and the Americans when the former launched the world's first artificial satellite, Sputnik 1 in 1957. The United States established the National Aeronautics and Space Administration (NASA) a year later and galvanized itself into action.

Space exploration has always been, and will be driven by the need for political and technological one-upmanship and, as Neil de Grasse Tyson, Director of the Hayden Planetarium at the American Museum of Natural History says: "If the United States commits to the goal of reaching Mars, it will almost certainly do so in reaction to the progress of other nations—as was the case with NASA, the Apollo programme, and the project that became the International Space Station. For the past decade, I have joked with colleagues that the United States would land astronauts on Mars in a year or two if only the Chinese would leak a memo that revealed plans to build military bases there".[83] Tyson goes on to say that this joke should not be taken lightly as the Chinese have released an official strategy paper in which they claim that they have a 5-year plan to advance their space capabilities which include the launching of space laboratories, manned spaceships and space freighters and engaging in other activities of advanced space exploration.

Although the Space Shuttle has been retired, NASA continues with its space programme, if only for the incontrovertible fact that slowing down space exploration would be disastrous for the development of humankind. Most of modern technology such as magnetic resonance imaging (MRIs) and other scanners have been developed by physicists. An example is the Hubble Space Telescope, which had serious optics defects when it was launched in 1991, prompting physicists to correct the deficiency and enable the telescope to send some of most sophisticated imagery from space to Earth. During this corrective process, scientists discovered that the challenges faced by astrophysicists in correcting Hubble's imaging problems were similar to what doctors faced in their visual search for tumours in mammograms. A collaborative effort which followed between the medical profession and astrophysicists enabled medical researchers to apply the corrective principles of the Hubble imagery to mammography, resulting in significant advances in the early detection of breast cancer.

The most recent initiative in space exploration is mining near earth asteroids (NEAs) where the presence of humans in the vicinity of NEAs brought to near earth orbit is envisioned for 2015.

---

[82] Neil de Grasse Tyson, The Case for Space—Why We Should Keep Reaching for the Stars, *Foreign Affairs*, March/April 2012, 22–33 at p. 23.

[83] *Ibid.*

On 18 April 2012 it was reported that a new company—Planetary Resources—announced its existence. It is supported by high end industrialists such as Google co-founders Larry Page and Sergey Brin, award winning film maker James Cameron and other persons of substance and wealth. The company is reported to have stated that it will overlay two critical sectors—space exploration and natural resources that would add trillions of dollars to the global GDP.[84]

A recent study (Asteroid Retrieval Feasibility Study) conducted in September 2011[85] and released in April 2012 by the Keck Institute of Space Studies (KISS) of the California Institute of Technology (CALTEC) opines that by the use of a three stage technological process: identifying candidate asteroids for mining (which has already been mastered); applying solar electric propulsion (which is now being used on small spacecraft); and having a human presence in the area in which an asteroid is directed (achievable in 2025),[86] a robotic asteroid retrieval process could be initiated, where aNEA of around 500 metric tons with a 7 m width could be "nudged" into close orbit with Earth. So far, 6 Apollo missions have brought back only 382 kg of samples.

The benefits of mining asteroids are expected to be substantial in terms of natural resources that can be brought back to Earth, estimated to add trillions of dollars to the global GDP. Among the natural resources expected are platinum and cobalt which are of high value and limited quantity here.

The next step would be to consider how this could be politically achieved, given the realm of space law that is applicable. Article 1 of the Outer Space Treaty[87] provides that the exploration and use of outer space, including the moon and other celestial bodies, shall be carried out for the benefit and in the interests of all countries, irrespective of their degree of economic or scientific development, and shall be the province of all mankind. It goes on to say that outer space, including the moon and other celestial bodies, shall be free for exploration and use by all States without discrimination of any kind, on a basis of equality and in accordance with international law, and there shall be free access to all areas of celestial bodies.

Finally, Article 1 provides that there shall be freedom of scientific investigation in outer space, including the moon and other celestial bodies, and States shall facilitate and encourage international co-operation in such investigation.

---

[84] http://www.technologyreview.com/blog/mimssbits/27776/ Within the next decade, the company hopes to use robots to prospect asteroids and gather rare earth materials, which are vital to medical devices, hand-held electronics and computers. Separately, Planetary Resources of Seattle wants to help NASA astronauts on deep-space missions to Mars. See http://www.standard.net/stories/2012/04/24/company-sets-course-mine-asteroids.

[85] Asteroid Retrieval feasibility Study, Prepared for the Keck Institute of Space Studies, 2 April 2012.

[86] *Id.* at p. 8.

[87] Treaty on Principles Governing the Activities of States in the Exploration and Use of Outer Space, including the Moon and Other Celestial Bodies), opened for signature at Moscow, London and Washington on 27 January 1967, 610 UNTS 205.

The more challenging provision in the Treaty is Article 2 which prescribes that outer space, including the moon and other celestial bodies, is not subject to national appropriation by claim of sovereignty, by means of use or occupation, or by any other means. This precludes a State from appropriating a celestial body inter alia by use.

The KECK Study has the following recommendations:

> The retrieval of a several-hundred-ton carbonaceous asteroid would present unparalleled opportunities for international cooperation. The retrieval could be carried out under the same philosophy as the Apollo program, "in peace for all mankind," but with a significant advantage. An international panel could be formed to oversee both curation of the body and the review of proposals for its study. The demand for samples for engineering and scientific study of the carbonaceous chondrite material by academic, governmental, and industrial laboratories—usually severely hampered by lack of pristine material—could be met generously. Samples could be returned to Earth for study, whereas microgravity processing experiments of the sort envisioned above could be carried out *in situ* in its parking orbit. Selected space faring nations would have access to the body under the oversight of the international curatorial panel. Nations without the ability to fly missions to the body would be encouraged to form teaming arrangements and propose jointly with those who can access it.
>
> As a natural step in moving human exploration capabilities from the International Space Station (ISS) into cislunar space, then beyond, the ACR mission concept would offer many opportunities for international participation.[88]

The two critical questions in this equation are: "if private companies or individuals conduct outer space activities, who is responsible and accountable for such activities?" and "who would own the samples brought to earth? If trillions of dollars worth of platinum and cobalt (not to mention other precious resources) are mined from asteroids nudged to near earth orbit, could there be private ownership of this property? In response to the first question on responsibility, The Outer Space Treaty provides that States Parties bear international responsibility for national activities in outer space, including the moon and other celestial bodies, whether such activities are carried out by governmental agencies or non-governmental entities, and for assuring that national activities are carried out in conformity with the provisions set forth in the treaty. The Treaty further states that the activities of non-governmental entities in outer space, including the moon and other celestial bodies would in essence inevitably require authorization and continuing supervision by the appropriate State Party.

As for the second question of ownership, the Outer Space Treaty prescribes that outer space is a resource that belongs to all mankind.[89] Therefore it follows that property which resides in outer space belongs to all. However, economic theory suggests that property rights and claims thereto emerge when it is in someone's self interest to claim property, and that claims to such rights are prompted by desires of

---

[88] *Supra*, note 201 at p. 10.

[89] The extent of State sovereignty over outer space and its component parts is prohibited by international space law, with freedom of space being its main principle. See Vershchetin 1977, pp. 429–436 at 436.

States, governments or individuals purely based on cost benefit possibilities. In this context, one has to wait and see what will develop in this expensive but valuable exercise worth trillions of dollars.

4.3.7.1 International Collaboration

The recommendation of the KECK Study—that an international panel be formed to oversee both curation of the body and the review of proposals for its study—is not without precedent. In 1965 Intercosmos, a multilateral programme of cooperation in outer space exploration was formed comprising nine socialist States.[90] These States, through various working groups, set up the following principles: each country has the right to declare its interest in any question examined by the working group and working groups must ensure that their decisions and recommendations are acceptable to all the nine countries (failing unanimous agreement the decisions are binding only on the States which supported their adoption and a State, which declares that it is not interested in a particular issue can always address it at a later stage); and the procedure of work and decisions taken did not impinge upon the sovereignty of the States concerned.[91]

In April 2007, 14 space agencies[92] put forward a Global Exploration Strategy (GES) and a Framework for Cooperation[93] which called for the development of an international exploration coordination tool to enhance mutual understanding among partners and to identify areas for potential cooperation. The GES advised that by jointly creating a common language of exploration building blocks, planners and engineers will be able to agree how practical features such as communications, control, life support and docking systems could be made to work together. The result would be that the 'interoperability' advocated by the GES between space vehicles will lower the risks of space exploration and could assure crew safety in case of life-threatening emergencies. The overall philosophy of the GES was that:

> Global-scale space exploration represents the sum of many projects undertaken nationally and internationally. But it also signifies a collective will to find answers to profound scientific questions, to create new economic opportunity and to expand the boundaries of

---

[90] Bulgaria, Hungary, German Democratic Republic, Cuba, Mongolian Peoples' Republic, Poland, Romania the Soviet Union and Czechoslovakia.

[91] For a more detailed discussion on Intercosmos, See Vereshchetin 1976, pp. 243–254.

[92] "Space agencies" refers to government representatives that include space agencies, science organizations and groups of space agencies that have been designated by their government to represent them. The 14 Agencies were In alphabetical order: ASI (Italy), BNSC (United Kingdom), CNES (France), CNSA (China), CSA (Canada), CSIRO (Australia), DLR (Germany), ESA (European Space Agency), ISRO (India), JAXA (Japan), KARI (Republic of Korea), NASA (United States of America), NSAU (Ukraine), Roscosmos (Russia). "Space Agencies" refers to government organizations responsible for space activities.

[93] The Global Exploration Strategy—The Framework for Cooperation: April 2007.

human life beyond Earth. These goals of space exploration in the service of society are embodied in the recurring themes of the Global Exploration Strategy.[94]

International collaboration[95] must be initiated by at least one player and achieves many objectives which include boosting resources for exploration; achievement of international diplomacy and comity; and obviating the duplication of efforts in space exploration between States. Arguably the most salient issue in international collaboration is that since principles of space law have States as the focus (as against private entities) the interests of States involved in such collaboration are paramount, in particular the initiating State. Space exploration has the preeminent objective of benefit to the initiating State which will reserve for it certain prerogatives and controls while ensuring that participating States and space agencies also retain flexibility of involvement and independence in key areas, including the undertaking of missions and carrying out key projects. In 1996, the United Nations General Assembly adopted a Declaration under the auspices of the United Nations Office for Outer Space Affairs which provides inter alia that in the exploration and use of outer space, States be guided by the principle of co-operation and mutual assistance and conduct all their activities in outer space with due regard for the corresponding interests of other States. If a State has reason to believe that an outer space activity or experiment planned by it or its nationals would cause potentially harmful interference with activities of other States in the peaceful exploration and use of outer space, it is required to undertake appropriate international consultations before proceeding with any such activity or experiment. A State which has reason to believe that an outer space activity or experiment planned by another State would cause potentially harmful interference with activities in the peaceful exploration and use of outer space may request consultation concerning the activity or experiment.[96]

4.3.7.2 Registration and Liability

Article VI of the *Outer Space Treaty* provides in part that State Parties to the Treaty bear international responsibility for national activities in outer space, whether such activities are carried out by governmental agencies or non-governmental agencies. This provision clearly introduces the notion of strict liability *ergaomnes* to the application of the *jus cogens* principle relating to outer space activities of States and could be considered applicable in instances where States hold out to the

---

[94] *Id.* Chapter 2, p. 7.

[95] "International collaboration", for the purposes of this article means all types of interactions between space agencies or States.

[96] Resolution 51/122 of 13 December 1996. See also General Assembly Resolution 110 (II) of 3 November 1947, which condemned propaganda designed or likely to provoke or encourage any threat to the peace, breach of the peace, or act of aggression, and considering that the aforementioned resolution is applicable to outer space, and Resolutions 1721 (XVI) of 20 December 1961 and 1802 (XVII) of 14 December 1962, adopted unanimously by the States Members of the United Nations.

international community as providers of technology achieved and used by them in outer space, which is used for purposes of air navigation. Article VI further requires that the activities of non-governmental entities in outer space shall require authorization and continuing supervision by the appropriate State Party to the Treaty, thus ensuring that the State whose nationality the entity bears would be vicariously answerable for the activities of that organization, thereby imputing liability to the State concerned.

Article VII makes a State Party internationally liable to another State Party for damage caused by a space object launched by that State. The *Registration Convention* of 1974[97] in Article II(1) requires a launching State of a space object that is launched into earth orbit or beyond, to register such space object by means of an entry in an appropriate registry which it shall maintains and inform the Secretary General of the United Nations of the establishment of such a registry. This provision ensures that the international community is kept aware of which State is responsible for which space object and enables the United Nations to observe outer space activities of States. Article VI of the Convention makes it an obligation of all State Parties, including those that possess space monitoring and tracking facilities, to render assistance in identifying a space object which causes damage to other space objects or persons. Justice Manfred Lachs analysed these provisions of the *Registration Convention* to mean that the State of registry and the location of the space object would govern jurisdictional issues arising out of the legal status of space objects.[98] On the issue of joint launching of space objects, Justice Lachs observed:

> No difficulties arise whenever a State launches its own object from its own territory; the same applies to objects owned or launched by non-governmental agencies registered in that State. However, in cases of joint launching, agreement between the parties is required as to which of them is to be deemed the "State of Registry". A similar agreement is also necessary when a launching is carried out by an international organization.[99]

The above provision ensures the identification of parties responsible for specific activities in outer space and thereby makes it easier to impose liability for environmental damage caused.

The Outer Space Treaty,[100] while expostulating the fundamental principle in its Article 1 that the exploration and use of outer space, including the moon and other celestial bodies, shall be carried out for the benefit and in the interests of all countries, explicitly imposes in Article VII international liability and responsibility on each State Party to the Treaty, for damage caused to another State Party or to its

---

[97] *Convention on Registration of Objects Launched into Outer Space*, adopted by the General Assembly of the United Nations, New York, 12 November 1974, 1023 UNTS 15.

[98] Lachs 1972, p. 70.

[99] *Ibid.*

[100] Treaty on Principles Governing the Activities of States in the Exploration and Use of Outer Space, including the Moon and Other Celestial Bodies, opened for signature at Moscow, London and Washington, 27 January 1967, 610 UNTS 205.

populace (whether national or juridical) by the launch or procurement of launch of an object into outer space. In its preceding provisions the Treaty imposes international responsibility on States Parties for national activities conducted in outer space. The Treaty also requires its States Parties to be guided by the principle of co-operation and mutual assistance in the conduct of all their activities in outer space.[101] This overall principle is further elucidated in the same provision:

> States Parties to the Treaty shall pursue studies of outer space, including the moon and other celestial bodies, and conduct exploration of them so as to avoid harmful contamination and also adverse changes in the environment of the Earth resulting from the introduction of extra terrestrial matter.[102]

The *Moon Agreement*[103] of 1979 provides that in the exploration and use of the moon, States Parties shall take measures *inter alia* to avoid harmfully affecting the environment of the earth through the introduction of extra terrestrial matter or otherwise.[104]

The *Liability Convention*[105] contains a provision which lays down the legal remedy in instances of damage caused by Space objects. Article II provides:

> A launching State shall be absolutely liable to pay compensation for damage caused by its space objects on the surface of the Earth or to aircraft in flight.[106]

Thereby imposing a regime of absolute liability on the State that launches space objects, from satellites which provide technology and communication that is used for air navigational purposes, to spacecraft carrying passengers. Although admittedly, both the *Outer Space Treaty* and the *Liability Convention* do not explicitly provide for damage caused by technology and communication provided by space objects, culpability arising from the "common interest" principle and liability provisions of the two conventions can be imputed to States under these Conventions.

Gorove states that in the field of international space law, two clearly connected terms have been used: liability and responsibility.[107] Although "responsibility" has not been cohesively interpreted in any legal treaty relating to outer space, "liability" occurs in the *Liability Convention* and is sufficiently clear therein. This, however, does not mean that State responsibility is not relevant to the obligations of States as,

---

[101]   *Id.* Article IX.

[102]   *Ibid.*

[103]   Agreement Governing the Activities of States on the Moon and other Celestial Bodies, signed on 5 December 1979, UN Doc A/RES/34/68 of 5/12/1979.

[104]   *Id.* Article 7.

[105]   *Convention on International Liability for Damage Caused by Space Objects*, March 29 1972, 24 U.S.T 2389, T.I.A.S No. 7762.

[106]   Article II(a) defines damage as including loss of life, personal, injury or other impairment of health; or loss or damage to property of States or of persons natural or juridical, or property of international governmental organizations.

[107]   Gorove 1983, p. 373.

in international relations, the invasion of a right or other legal interest of one subject of the law by another inevitably creates legal responsibility.

4.3.7.3 Policy Issues

It is a truism that no individual country or group of countries can regulate outer space, which is governed by a patchwork of international treaties, resolutions of the United Nations and industry standards. However, international relations and domestic policy drive a nation's direction towards outer space exploration and reflect individual State interests. The United States, which incontrovertibly is the leader among all space faring nations (which include the BRIC States—Brazil, Russia, India and China—and France, Japan, the United Kingdom, Israel the European Space Agency,[108] Ukraine and Iran),[109] is responsible for 75 % of space funding worldwide and owns or operates 40 % of all active spacecraft in orbit.[110]

Garold Larson, Alternate Representative to the First Committee of the 64th Session of the United Nations Assembly held on 19 October 2009, succinctly outlined the policy of the United States on space exploration. The foremost principle outlined by Larson was that the United States will continue to uphold the principles of the 1967 Outer Space Treaty, which the United States recognized as providing fundamental guidelines required for the free access to and use of outer space by all nations for peaceful purposes.[111] He went on to say that the United States will continue to take an active role in identifying and implementing cooperative efforts with established and emerging members of the international space-faring community to ensure the safety of the space assets of all nations and also

---

[108] The European Space Agency (ESA) is Europe's gateway to space. Its mission is to shape the development of Europe's space capability and ensure that investment in space continues to deliver benefits to the citizens of Europe and the world.

[109] With regard to other States, Canada continues to develop its contributions to science experiments for the International Space Station, the world's largest microgravity laboratory. The Special Purpose Dexterous Manipulator (Dextre), one of the Canadian robotic contributions to the station, is now being prepared for launch. On 17 July 2001, The Chilean legislature issued Supreme Decree No. 338, creating a Presidential Advisory Committee known as the Chilean Space Agency and signed by the President of the Republic, the Secretary-General of the Presidency and the Ministers for Foreign Affairs, National Defence, Finance, Transport and Telecommunications and Education. The Agency inter alia advises the President of the Republic on all matters concerning the identification, formulation and implementation of policies, plans, programmes, measures and other activities relating to space; and propose national space policy and the measures, plans and programmes required for the implementation thereof. Since the 1970s, Morocco has gradually incorporated outer space into its development programmes. The firm commitment of the Government of Morocco has enabled the country to make significant progress in space telecommunications, space meteorology and remote sensing.

[110] http://www.cfr.org/space/code-conduct-outer-space/p26556.

[111] Statement by Garold N. Larson, Alternate Representative to the First Committee, on Outer Space (Disarmament Aspects), in the First Committee of the Sixty fourth Session of the United Nations General Assembly, at http://usun.state.gov/briefing/statements/2009/130701.htm.

expand cooperation with other like-minded spacefaring nations and with the private sector to identify and protect against intentional and unintentional threats to its space capabilities.[112]

The European Union, in 2008, published a draft Code of Conduct for Outer Space Activities, which it later revised in September 2010. The fundamental postulate of this code is that member states should establish policies and procedures to minimize the possibility of accidents… or any form of harmful interference with other States' right to the peaceful exploration and use of outer space. The Code applies three basic principles in pursuance of its overall objective: freedom of access to space for peaceful purposes; preservation of the security and integrity of space objects in orbit; and due consideration for the legitimate defence interests of states. The code is not a legislative instrument and therefore has no legally binding effect on member States. It remains a voluntary agreement among states with no formal enforcement mechanisms. On 4 April 2011 the European Commission published a space strategy for Europe whereby the European Union seeks to identify and support the development of essential technologies for exploration, in particular in the fields of energy, health and recycling (support for life in isolated environments). These matters are not necessarily dealt with in the space sector itself and cross-fertilisation should be promoted with other sectors in order to benefit the citizens directly.

China's space programme, which started in the 1950 as part of its nuclear weapons programme, burgeoned steadily to enable China the third country, after the United States and Russia, to continue with a sustained manned space programme.[113] In view of China's exclusion from the International Space Station programme China accelerated its own space programme and, in a statement issued in 2010 stated;

> In accordance with the principle of peaceful use of outer space, China has conducted bilateral cooperation and exchanges with Russia, France, Brazil, Ukraine, the United States and the European Space Agency (ESA) in the fields of space technology, space exploration and space science. It supports the work of the United Nations Committee on the Peaceful Uses of Outer Space (COPUOS) and Asia-Pacific Space Cooperation Organization (APSCO), and plays an active role in making use of outer space technologies to conduct multilateral cooperation in Earth science research, disaster prevention and reduction, deep space exploration, and space debris mitigation and protection.[114]

China, through its space programme, seeks to achieve military capability and diplomatic acceptance and credibility. A commentator expresses Chinese space

---

[112] *Ibid.*

[113] The US spends approximately seven times as much as China annually on its space program, $16 billion compared to 2.2 billion, (34) but China's costs are lower. Rosita Dellios, China's space program: A strategic and political analysis, *Culture Mandala: The Bulletin of the Centre for East-West Cultural and Economic Studies* 7.1 (2005). Also at http://works.bepress.com/rosita_dellios/24/.

[114] State Council Information Office:2010, reported in Rosita Dellios, China and Outer Space, Bond Universirty e-publication at http://works.bepress.com/rosita_dellios/49/.

policy in Chinese dialectical terms eloquently, that 'it represents a distinctive space policy in which the yin of Active Defence secures the prospects for the yang of cooperative governance which, in turn, promotes a common security".[115]

In Space Law, there is no such being as a "person" in outer space. There are only astronauts and personnel.[116] The 1967 Outer Space Treaty stipulates that State parties to the Treaty shall regard astronauts as envoys of mankind in outer space and shall render to them all possible assistance in the event of accident, distress or emergency landing on the territory of another State party or on the high seas.[117] The provision also requires State parties to return astronauts under the above circumstances safely and promptly to the State of registry of their space vehicle.[118]

The Treaty provision is a reproduction *verbatim* of Para 9 of United Nations General Assembly Resolution XVIII of 1962.[119] Although initially, the world's "envoys of mankind" seemingly created some apprehension in the international community as to whether such phraseology connoted diplomatic immunity to astronauts, Bin Cheng clears up this ambivalence by concluding that it was only a figure of speech which has not been repeated in any United Nation's documentation yet.[120] The perceived inadequacy of definitive identification at international law of an astronaut and his conduct in outer space leaves one with the basic premise that international law is incontrovertibly applicable to outer space activities and outer space, including the moon and other celestial bodies, which are totally independent of appropriation by States or individuals. This in turn leaves one with the inevitable conclusion that outer space would be analogous at international law to the high seas.

Jurisdictionally, any person comes clearly within the purview of the State on whose territory he is or above whose territory and in the airspace of the State concerned, if he is in an aircraft. Generally, in outer space, this *status quo* changes, and astronauts become liable under the laws of the State of registry or the State which launches their spacecraft for travel and work in outer space.[121] This is brought to bear by Article VIII of the Outer Space Treaty which provides:

---

[115] Dellios, *Id.* at 8.

[116] Treaty on Principles Governing the Activities of States in the Exploration and Use of Outer Space, Including the Moon and Other Celestial Bodies, Opened for Signature at Moscow, London and Washington on 27 January 1967, 610VNTS 205. It must be noted that the first "space tourist" Denis Tito was called a space tourist for purposes of public reference by the media. He was called a "guest cosmonaut" by the Russians and an amateur astronaut by the Americans. See http://www.spacedaily.com/news/011206133411.3i4zwq28.html.

[117] *Id.* Article V.

[118] *Ibid.*

[119] UNGA Resolution 1962 (XVIII) Declaration of Legal Principles Governing the Activities of States in the Exploration and Use of Outer Space.

[120] Cheng 1997, pp. 259 & 460.

[121] *Id.* 458.

> A State party to the Treaty on whose registry an object launched into outer space is carried shall retain jurisdiction and control over such object, and over any personnel thereof, while in outer space or on a celestial body.[122]

However, as Bin Cheng validly points out[123] the interpretation of Article VIII could well result in ambivalence and confusion. The "object" and "personnel" referred to in the Treaty provision do not adequately cover persons who are not "personnel" such as passengers in a spacecraft. Of course, as Cheng maintains, the quasi jurisdiction of the State of registry of the spacecraft can apply both in the instance of conduct in the spacecraft as well as outside the spacecraft on the basis that the astronaut concerned would be deemed to belong to the spacecraft at all times in outer space. Logically, therefore, such jurisdiction could be imputed to passengers, visitors and guests by linking them to the spacecraft in which they travelled. This far reaching generalization would then cover the conduct of an astronaut or other persons while walking on the moon, Mars or other celestial body, as well as such persons who go on space walks outside the spacecraft in which they travelled.

Another provision which sheds some light on past attempts by the international community to identify liability and jurisdictional issues relating to astronauts is Article 12 of the Moon Treaty of 1979[124] which provides:

> States Parties shall retain jurisdiction and control over their personnel, space vehicles, equipment facilities, stations and installations on the moon...

It is presumed that the legal link between the personnel and the spacecraft they travel in under the circumstances are imputed to the State of registry of the said craft. If this were not the case, and such a link cannot be established, the provision itself becomes meaningless and destitute of effect.

The above provisions, although seemingly adequate for an incipient world space programme, do not adequately address modern exigencies of outer space activity such as collaboration in space stations where repair missions and salvage activities may call for multinational crews, joint space exploration calling for multiple space technology, and transportation to outer space of passengers.

4.3.7.4 Conduct of the Person in Outer Space

The recognition that the scope of manned space flight is being expanded from the flight of astronauts to other persons such as repair crew and passengers, is becoming

---

[122] Like the earlier cited Treaty provision, this provision is derived from United Nations documentation and has been reproduced almost verbatim from para 7 of the 1963 General Assembly Declaration appearing in Resolution 1962 (XVIII). The Treaty provision extends the scope of application of the provision to conduct of astronauts both inside and outside the spacecraft.

[123] Bin Cheng, *supra* note 236 at p. 459.

[124] Agreement Governing the Activities of States on the Moon and other Celestial Bodies, UN Doc A/RES/34/68 of 5 December 1979.

evident. In an attempt in 1988 at drafting a Convention on manned space flight, a team of draftsmen comprising a distinguished cluster of experts in space law from Germany, the Russian Federation and the United States succeeded in a sustained attempt at producing a draft legal instrument which covers certain exigencies of personal conduct in space travel. The most significant thrust of this draft Convention is that it blends harmoniously the essential qualities of scholarship and practicality. The draft Convention has been published in order to draw the attention of the world community of space lawyers and seek comments. It effectively conveys the fundamental postulate that manned space flight is the cornerstone of exploration of outer space and therefore its development requires guidelines on international cooperation and liability.

The draft Convention, in Article III links itself to the Outer Space Treaty principle of awarding jurisdiction in relation to a manned space object and person therein to the State of registry in relation to occurrences in outer space or in a celestial body or on or in the high seas or any other place beyond the jurisdiction of any State. Article IV of the draft Convention devolves responsibility and authority over a manned space flight, the space object involved in such flight and all persons on board, on the commander of the space object. The commander is given sole authority throughout the flight to use any reasonable and necessary means to achieve this end. The same provision makes both the commander and all members of the crew answerable to a person identified as the Director of Manned Space Flight Operations, who is defined by the draft Convention as a person who is designated by the State exercising jurisdiction and control over the space object to be in charge of a particular manned space flight. By this measure, the draft Convention skillfully and unequivocally identifies the chain of command, giving the commander absolute authority on the spot over all those in the space object during the flight, while making him answerable to a person designated as Director of Manned Space Flight Operations, who presumably will b e on ground and in mission control.

The draft Convention also ensures safety of persons involved in a manned space flight whether they be crew, passengers or any other category of persons affected by such flight. One of the strengths of the draft Convention is its provision with regard to environmental pollution or other harm caused to the environment by a manned space flight, where the instrument lays responsibility on States whose manned space exploration may jeopardise an existing environmental balance. It also provides for assistance to be given by persons in a manned space flight to others in distress in outer space and prescribes international responsibility on States, whether the space flights in question are carried out by governmental or non governmental entities.

Outer space and celestial bodies can be used for the common heritage of mankind but are *res extra commercium* like the high seas. However, here the distinction ends, in that unlike the high seas, which can be appropriated in certain circumstances, such as through acquiescence by one State of appropriation of an area of the high seas by another, outer space or celestial bodies cannot be appropriated under any circumstances. It is not possible to apply the principle of appropriation to the conduct of crews of other persons in outer space. One cannot establish a pattern of conduct as a prescriptive right in outer space because there are no territorial limits demarcated by and between individual States in outer space. *A fortiori*, outer space has been identified as one composite area which cannot be appropriated by one particular State to the exclusion of others by Treaty provision.

Freedom of outer space, which lays the foundation for conduct of persons in outer space, is enshrined in Article 1 of the Outer Space Treaty of 1967, which stipulates that the exploration and use of outer space, including the moon and other celestial bodies, shall be carried out for the benefit and in the interests of all countries, irrespective of their degree of economic or scientific development, and shall be the province of all mankind. The provision also requires outer space to be free for exploration and use by all States without discrimination of any kind, on a basis of equality and in accordance with international law. Finally, the provision grants free access to all States in relation to all areas of celestial bodies.

On a purely superficial comparison of the freedoms of outer space and the high seas, one can notice a general similarity in that both areas are open to mankind equally. However, the purposes for which the areas can be used are intrinsically different. For example, space law is all encompassing on the subject of the conduct of humans in outer space and celestial bodies. The Outer Space Treaty makes the sweeping statement that outer space and celestial bodies shall be open for exploration by mankind, which includes, *inter alia*, the freedom to conduct research, experiments and other forms of exploration. The Convention on High Seas, on the other hand is inclusive and therefore restrictive in forming areas of specific activity. The Convention on the Law of the Sea of 1982 has somewhat remedied this *lacuna* by adding, in Article 87(1), the freedom of scientific research *inter alia* to the already existing four freedoms.

Criminal conduct is an area where the principle of international law applicable to the High Seas lend themselves as a useful analogy to space law. Of course, the offence of piracy cannot be committed by astronauts who are sent to outer space in spacecraft belonging to a State. The offence has to be committed for private ends by persons in a private ship or craft. The offence of piracy in the high seas would nonetheless apply as an analogy to a similar offence committed by private individuals in outer space who do not represent a State as official crew members. This would cover the improbably but nonetheless possible events of the future such as a mutiny on board a commercial spacecraft carrying passengers (which is an analogy derived from shipping law). Piracy in outer space may also occur in instances where personnel of a space craft could act on the orders of a recognized government which is in gross breach of international law and which show a criminal disregard for human life.

Of direct analogy to the draft Convention are three Conventions and a Declaration at air law which lay down principles of law with regard to criminal conduct of persons on board aircraft. Of course, understandably, given the fact that crimes on board aircraft have been prolific over the past years in comparison, these instruments are somewhat more evolved than the space law initiative and have concentrated on such offences as hijacking and other terrorist acts. But they also provide for sanctions against general criminal conduct and therefore would form a useful paradigm for future space law initiatives, particularly on the subject of terrorism in outer space and celestial bodies. The Convention on Offenses and Certain Other Acts Committed on Board Aircraft (1963),[125] refers to any offence committed or act done by a person on board any aircraft registered in a contracting State, while the aircraft is in flight or on the surface of the high seas or of any other area outside the territory of such State.[126] The aircraft is considered to be in flight from the moment power is applied for the purpose of take off until the moment when the landing run ends.[127] In addition, the Tokyo Convention mentions acts of interference, seizure of or other wrongful exercise of control of an aircraft, implying its concern over hijacking.[128]

The Hague Convention of 1970[129] in Article 1 identifies any person who, on board an aircraft in flight, unlawfully by force or threat or by any other form of intimidation seizes or takes control of such aircraft, or even attempts to perform such an act, as an offender.[130] Anyone who aids such an act is an accomplice, and is included in the category of the former.[131]

It is clear that the Hague Convention by this provision has neither deviated from Article 11 of the Tokyo Convention nor offered a clear definition of the offence of hijacking. It merely sets out the ingredients of the offence—the unlawful use of force, threat or any other form of intimidation and taking control of the aircraft. The use of physical force, weapons or firearms or the threat to use such modes of force are imputed to the offence in this provision. The words force, threat or intimidation indicate that the element of fear would be instilled in the victim. It is an interesting question whether these words would cover an instance where the use of fear as an implement to execute the offence of hijacking covers non-coercive measures such as the drugging of food or beverages taken by the passengers or crew. The Hague Convention does not ostensibly cover such instances. In this context, many recommendations have been made to extend the scope of its Article 1.[132] It is also interesting that the Convention does not envisage an instance where the offender is not on board the aircraft but remains on ground and directs operations therefrom after planting a dangerous object in the aircraft. According to Article 1, the offence has to be devoid of a lawful basis albeit that the legality or illegality of an act is not clearly defined in the Convention. It is also a precondition in Article 1 that the offence has to be committed in flight, that is while all external doors of the aircraft

---

[125] Hereafter referred to as the Tokyo Convention.

[126] Tokyo Convention, Ch.1., Article 1(2).

[127] *Id.*, Chapter 1, Article 1(3).

[128] *Id.*, Chapter 4, Article 11.

[129] *Convention for the Suppression of Unlawful Seizure of Aircraft*, the Hague, 16 December 1970, hereafter referred to as the Hague Convention.

[130] *Id.*, Article 1(a).

[131] *Id.*, Article 1(b).

[132] See Shubber 1973, pp. 687 at 692.

are closed after the embarkation of the passengers and crew.[133] The mobility of the aircraft is immaterial. Furthermore, Article 1 is rendered destitute of effect if an offence is committed while the doors of the aircraft are left open.

The Tokyo Convention was perhaps the first major attempt at curtailing the menace of hijacking. Not only did it deal solely with jurisdiction over offenses committed in an aircraft in flight, but it also did not exclude any criminal jurisdiction which would have been exercised according to the provisions of any law.[134] The element of nationality underlines the parochial nature of the treatment of the offence and the obstinate refusal of the international community to infuse a universality to the treatment of the offence. Perhaps, as one commentator observed[135] the international community was not prepared in 1963 to address this problem on a collective basis.

The subsequent Hague Convention emphasises that each contracting State undertakes to impose severe penalties without defining what these penalties should be.[136] Furthermore, the geographic limitations set out in Article 1 curtails the punitive measures recommended in the Convention significantly. The Convention makes a further serious omission in stating that it applies only if the place of take off and place of landing is outside the State of registration of the aircraft.[137] This gives rise to a serious anomaly in that if an aircraft with a destination outside its territory of nationality is seized in mid air prior to leaving its airspace and brought back to the place of take off the Convention would not apply even to scheduled flight.

The Convention for the Suppression of Unlawful Acts Against The Safety of Civil Aviation signed at Montreal on 23 September 1971,[138] although extending the period in which the offence would be committed to a period beginning at pre flight preparation which ends 24 h after landing, does not cover acts of sabotage, destruction or any damage effected before the above period starts after 24 h after landing. The Montreal Convention however, is the best attempt so far at attempting to control or curb the offence of hijacking on an international level. Even so, the three attempts so far at international accord fail to cover certain gaps which still exist in this area of prevention and control of hijacking. They are:

(a) That the Conventions do not provide for and guarantee the trial of an offender and do not specify adequate punitive measures;
(b) That no obligation is cast on contracting States for the extradition of an offender;
(c) That no provision is made for the universal adoption of standards of precaution and safety; and,

---

[133] Hague Convention, *supra,* note 129 Article 3, para 1.

[134] Tokyo Convention, *op.cit.,*Article 3(3).

[135] A.F. Lowenfeld, *Aviation Law, Cases and Materials*, M. Bender, New York, 1972, at p. 87.

[136] Hague Convention, *op.cit.*, Article 2.

[137] *Id.* Articles 1(4), 3, 4, and 7.

[138] Hereafter referred to as the Montreal Convention.

(d) That the initial attempt, *albeit* somewhat unsophisticated, of the Tokyo Convention at a remedial approach has been thwarted by the repressive attitude of the two subsequent Conventions.

The Bonn Declaration of 1978 was yet another attempt by the international community to combat terrorism related to international civil aviation. The major economic powers Canada, France, Federal Republic of Germany (as it then was), Italy, Japan, UK and USA collaborated to intensify efforts to combat terrorism. The seven signatory States pledged to take immediate action to cease all flights to a country which refuses to extradite hijackers or return hijacked aircraft, and to halt all incoming aircraft from that country or from any airlines of that country.

The Montreal Convention fails to define in specific terms, the offence of hijacking, although it circumvents barriers placed by Article 1 of the Hague Convention.[139] For instance : it encircles instances where an offender need not be physically present in an aircraft; includes instances where an aircraft is immobile; its doors open; and even draws into its net any person who disseminates false information which could endanger an aircraft in flight.[140] None of the three conventions have however succeeded in identifying the offence of hijacking or advocating preventive measures against the offence itself.

Two issues emerge from the discussions in this article. Firstly, that space exploration must continue for technology to progress. Mining asteroids is a promising element in this equation that would pave the way for advances in physical and medical science as well as chemistry. Above all it will give humankind a sense of perspective, as to who we are, where we have come from, and where we are headed. Secondly, since space diplomacy is an incipient but rapidly evolving process, the key to international cooperation would lie in relations between the United States and China. Both the countries have, in their policies, recognized the preeminent principle—that space exploration should be for the benefit of all humankind. This is a good starting point. Mining asteroids is mainly for scientific advancement and both countries could focus their relations on cooperative governance towards the achievement of that goal. China would feel included and identify itself away from its cosmic isolation. The United States would achieve its vision of international cooperation and both countries could consolidate their ambitions of scientific advancement and values.

A joint space programme between key players of North America, Europe and Asia could greatly stabilize Asia and very likely forge reconciliation between China and Japan and obviate burgeoning rivalry between China and India.[141] As for China's position in international cooperation in outer space affairs, Wang Xiaoying, Secretary-General of China Arms Control and Disarmament Association has stated that in the first 10 years of the 21st century, China progressed well in space

---

[139] Montreal Convention, Article 1.

[140] *Ibid.*

[141] Zbigniew Brzezinski, Balancing the East, Upgrading the West, *Foreign Affairs*, January/February 2012, 97–104 at p. 97.

exploration which by launching manned space flights, performing spacewalk, establishing the Beidou navigation system, and conducting anti-satellite and anti-ballistic missile tests. The Secretary General ventured to say that, owing to the financial crisis during this decade experienced by the United States, it was forced to restrict the development of its outer space technology and end its space shuttle program, which was a major setback for US space technology development.

Xiaoying has also said that it is in the interest of all countries to establish a new framework of international laws on and norms for the use of outer space, recognizing that the fact that the United States has indicated an interest in establishing a dialogue with China on space shows that the United States is concerned about the direction and intention of China's space strategy. Encouragingly, the Secretary General has asserted that China should understand the concern of the United States and respond openly and confidently, concluding that it has been China's constant endeavour to oppose confrontation and have dialogue, provided such dialogue is founded on equality and mutual respect.[142]

Given this scenario, and the fact that both countries have adopted what they call a "constructive partnership" in world affairs, the United States could, with the association of a strong Europe and Russia, engage in inclusive discussions with China on collaborative involvement in space exploration. These diplomatic necessities do not detract from the fact that there has to be a strict code of conduct adopted through international law in multilateral treaty form that would address the issue of criminality with regard to mined space property. Analogies of the law of the sea and air law could well be a starting point in this exercise.

References

Abeyratne RIR (1998) Negligence of the airline pilot. Prof Negligence 14(4):219–231

Becker (2006) Terrorism and the State; rethinking the rules of state responsibility. Hart Publishing, Portland

Blackstone, Morrison (eds) (2001) 4 Commentaries on the laws of England. Cavendish, London, p 68

Brownlie (1983) System of the law of nations: state responsibility, Part 1. Oxford Clarendon, Oxford, p 39

Caron (1998) The basis of responsibility: attribution and other trans-substantive rules. In: Lillich, Magraw (eds) The Iran-United States claims tribunal: its conclusions to state responsibility (Irvington-on–Hudson). Transnational Publishers, New York, vol 109, pp 153–154

Cheng B (1997a) Studies in international space law. Clarendon Press, Oxford, p 490

Cheng B (1997b) Studies in international space law. Clarendon Press, Oxford, pp 259 and 460

de Arechaga (1968) International responsibility. In: Sorenson M (ed) Manual of public International law. St. Martin's Press, New York, pp 531 at 535

De Vattel (tr) 2 (1916) The law of nations or, the principles of natural law: applied to the conduct and to the affairs of Nations and Sovereigns, Legal Classics Library, New York, p 72

---

[142] China Willing to Participate in Space Diplomacy, Wednesday August 3 2011. http://spaceports.blogspot.ca/2011/08/china-willing-to-participate-in-space.html.

Gorove S (1983) Liability in space law: an overview. Ann Air Space L VIII:373

Hyde (1928) Concerning damages arising from neglect to Prosecute 22. Am J Int L140 22:140–142

Lachs M (1972) The law of outer space, an experience in contemporary law making. Sijthoff, Leiden, p 70

Miyagi (2005) Serious accidents and human factors. American Institute of Aeronautics and Astronautics, Virginia, p 143

Shubber S (1973) Aircraft Hijacking under the hague convention 1970—a new regime? 22 Int Comp Law Q 687:692

Starke (1989) Introduction to International Law, 10th edn. Butterworths, London, p 3

Vereshchetin VS (1976) Intercosmos: present and future. Ann Air Space Law 1:243–254

Vershchetin (1977) On the principle of State Sovereignty in International space law. Ann Air Space Law II:429–436, 436

Warner K, Rooney K (1997) ICAO technical instructions provide a complete system for he transport of dangerous goods by air. ICAO J 52(2):23–24

Chapter 5
Some Preliminary Space Law Principles for Consideration by ICAO

5.1 Principles and Guidelines Contained in International Law

United Nations Resolution 1721 (XVI) established the primary principle that international law, including the Charter of the United Nations applies to outer space and celestial bodies and that outer space and celestial bodies are free for exploration and use by all States in conformity with international law and are not subject to national appropriation. This principle is also contained in Article 1 of the Outer Space Treaty of 1967.[1] It must be mentioned that this Resolution has no binding force on States and that it merely commends to States the aforesaid principles for their guidance. This notwithstanding, Resolution 1721 (XVI) was adopted unanimously by the General Assembly and is therefore supported by some States including the Russian Federation, United States and the United Kingdom. General Assembly Resolution 1962 (XVIII) contains the following principles in a Declaration, in addition to the guiding principles enunciated in Resolution 1721 (XVI):

(a) States bear international responsibility for national activities in outer space, whether carried on by governmental agencies or by non-governmental entities, and for assuring that national activities are carried on in conformity with the principles set forth in the Declaration. The activities of non-governmental entities in outer space shall require authorization and continuing supervision by the State concerned. When activities are carried on in outer space by an international organization, responsibility for compliance with the principles set forth in the Declaration shall be borne by the international organization and by the States participating in it;

---

[1] Treaty on Principles Governing the Activities of States in the Exploration and Use of Outer Space, Including the Moon and Other Celestial Bodies, signed at Washington, London, Moscow, January 27, 1967.

© The Author(s) 2015
R. Abeyratne, *Regulation of Commercial Space Transport*,
SpringerBriefs in Law, DOI 10.1007/978-3-319-12925-9_5

(b) In the exploration and use of outer space, States shall be guided by the principle of co-operation and mutual assistance and shall conduct all their activities in outer space with due regard for the corresponding interests of other States. If a State has reason to believe that an outer space activity or experiment planned by it or its nationals would cause potentially harmful interference with activities of other States in the peaceful exploration and use of outer space, it shall undertake appropriate international consultations before proceeding with any such activity or experiment. A State which has reason to believe that an outer space activity or experiment planned by another State would cause potentially harmful interference with activities in the peaceful exploration and use of outer space may request consultation concerning the activity or experiment;

(c) The State on whose registry an object launched into outer space is carried shall retain jurisdiction and control over such object, and any personnel thereon, while in outer space. Ownership of objects launched into outer space, and of their component parts, is not affected by their passage through outer space or by their return to the earth. Such objects or component parts found beyond the limits of the State of registry shall be returned to that State, which shall furnish identifying data upon request prior to return;

(d) Each State which launches or procures the launching of an object into outer space, and each State from whose territory or facility an object is launched, is internationally liable for damage to a foreign State or to its natural or juridical persons by such object or its component parts on the earth, in air space, or in outer space; and

(e) States shall regard astronauts as envoys of mankind in outer space, and shall render to them all possible assistance in the event of accident, distress, or emergency landing on the territory of a foreign State or on the high seas. Astronauts who make such a landing shall be safely and promptly returned to the State of registry of their space vehicle.

Another noteworthy instrument of space law is the *1968 Agreement* on the *Rescue* of *Astronauts* which addressed two issues, the first being the conduct of national manned space flight programmes where the Agreement recognises that accidents or mistakes may occur, and that astronauts may have to be rescued from space vehicles if they re-enter the earth's atmosphere from outer space and land somewhere outside the territory of the launching State. The second issue arises from manned and unmanned space programmes. It recognises that accidents or mistakes may occur and that as a result space objects or their component parts may re-enter the earth's atmosphere and land in areas outside the territory of the launching State. An interesting point has been raised by some commentators as follows:

Article 2 of the Assistance and Return Agreement require that when a landing has been made owing to accident, distress, emergency, or is unintended, the Contracting Party on whose territory the spacecraft has landed must "immediately take all possible steps to rescue" the personnel of the spacecraft and "render them all necessary assistance." Thus Article 2 requires, at least in theory, that the maximum possible rescue effort be made by the Contracting Party. It may be argued that the Contracting Party must utilize all resources available for the rescue effort, even to the point of diverting certain resources from other important uses. This appears to be a greater measure of assistance than that required under Article 25 of the Chicago Convention, which requires only that assistance deemed "practicable" by the state on which a landing is made; but it is consistent with the measure of "all possible assistance" required to be rendered to astronauts under Article V of the Outer Space Treaty. Whether a real distinction exists between "all possible" and. "practicable" will depend upon the efforts exerted by states in comparable situations.[2]

The 1968 Agreement goes on to say that If information is received or it is discovered that the personnel of a spacecraft have alighted on the high seas or in any other place not under the jurisdiction of any State, those Contracting Parties which are in a position to do so shall, if necessary, extend assistance in search and rescue operations for such personnel to assure their speedy rescue. They shall inform the launching authority and the Secretary-General of the United Nations of the steps they are taking and of their progress. If, owing to accident, distress, emergency or unintended landing, the personnel of a spacecraft land in territory under the jurisdiction of a Contracting Party or have been found on the high seas or in any other place not under the jurisdiction of any State, they will be safely and promptly returned to representatives of the launching authority. Furthermore the Agreement provides in Article 5 that each Contracting Party which receives information or discovers that a space object or its component parts has returned to Earth in territory under its jurisdiction or on the high seas or in any other place not under the jurisdiction of any State is required to notify the launching authority and the Secretary-General of the United Nations. Each Contracting Party having jurisdiction over the territory on which a space object or its component parts has been discovered is expected to,, upon the request of the launching authority and with assistance from that authority if requested, take such steps as it finds practicable to recover the object or component parts. Upon request of the launching authority, objects launched into outer space or their component parts found beyond the territorial limits of the launching authority will be returned to or held at the disposal of representatives of the launching authority, which shall, upon request, furnish identifying data prior to their return. A Contracting Party which has reason to believe that a space object or its component parts discovered in territory under its jurisdiction, or recovered by it elsewhere, is of a hazardous or deleterious nature may so notify the launching authority, which will immediately take effective steps, under the direction and control of the said Contracting party, to eliminate possible danger of harm. Expenses incurred in fulfilling obligations to recover and return a space object or its component parts will be borne by the launching authority.

In the context of liability, The *Moon Agreement* of 1979, provides that in the exploration and use of the moon, States Parties shall take measures *inter alia* to avoid harmfully affecting the environment of the earth through the introduction of extra terrestrial matter or otherwise. At the same time, the Moon agreement, which applies to other celestial bodies as well, in Article 6 provides that there shall be freedom of scientific investigation on the moon by all States Parties without

---

[2] Dembling and Daniel 1968, pp. 630–663 at 643.

discrimination of any kind, on the basis of equality and in accordance with international law. The same provision allows a State Party which finds minerals and other substances to collect and take custody of such material and even conduct scientific experiments on the samples. They are required to have regard to requests from other State Parties for the use of such material and also given the discretion to exchange scientific information in this context. In this context it must be noted that at the frontiers of this issue are the astronauts, who are by treaty designated as envoys of mankind in outer space, casting on them the responsibility of adhering to applicable treaty provisions on behalf of their States.

In the field of international space law, two clearly connected terms have been used: liability and responsibility. Although "responsibility" has not been cohesively interpreted in any legal treaty relating to outer space, "liability" occurs in the *Convention on International Liability for Damage Caused by Space Objects*, March 29 1972 (Liability Convention) and is sufficiently clear therein. This, however, does not mean that State responsibility is not relevant to the obligations of States law as, in international relations, the invasion of a right or other legal interest of one subject of the law by another inevitably creates legal responsibility.

At present, one can regard responsibility as a general principle of international law, a concomitant of substantive rules and of the supposition that acts and omissions may be categorized as illegal by reference to the rules establishing rights and duties. Shortly, the law of responsibility is concerned with the incidence and consequence of illegal acts, and particularly the payment of compensation for loss caused. Therefore, As discussed, both treaty law and general principles of international law on the subject of space law make the two elements of liability and responsibility a means to an end—that of awarding compensation to an aggrieved State or other subject under the law. In view of the many legal issues that may arise, the primary purpose of a regulatory body which sets standards on State liability would be to carefully consider the subtleties of responsibility and liability and explore their consequences on States and others involved as they apply to the overall concept of the status of a State as a user of space technology which may cause harm or injury to the latter.

We have to be mindful of a few fundamental truths. First, if we come across any form of life in outer space it will be the concern of all humankind. Second, any treatment of such life, irrespective of the fact that it is found in outer space, should be according to the principles of international law and the United Nations Charter, which contains numerous provisions which are relevant to the use of force. Several General Assembly resolutions, adopted without dissent or with near unanimity, have restated, amplified and clarified the meaning of these Charter provisions. Possibly, the most relevant and authoritative is the 1970 Declaration on Principles of International Law Concerning Friendly Relations and Co-operation Among States in Accordance with the Charter of the United Nations (Friendly Relations Declaration). Although some dispute exists in regard to the precise legal status of the *Friendly Relations Declaration*, it is generally regarded as an authoritative interpretation of broad principles of international law expressed in the Charter. Another, and more controversial example, is the General Assembly's "Definition of Aggression" resolution.

The UN Charter does not directly address the question of intervention by states; rather, under Article 2(7) it precludes the Organization itself from intervening "in matters which are essentially within the domestic jurisdiction of any state." Hence the General Assembly's Declaration on the Inadmissibility of Intervention in the Domestic Affairs of States and the Protection of Their Independence and Sovereignty, adopted in 1965 by a vote of 109 to none with one abstention, takes on special legal significance.

The United Nations General Assembly Resolution A/RES/51/122, which was adopted at the 83rd plenary meeting on 13 December 1996[3] provides that International cooperation in the exploration and use of outer space for peaceful purposes (hereinafter "international cooperation") shall be conducted in accordance with the provisions of international law, including the Charter of the United Nations and the Treaty on Principles Governing the Activities of States in the Exploration and Use of Outer Space, including the Moon and Other Celestial Bodies. It shall be carried out for the benefit and in the interest of all States, irrespective of their degree of economic, social or scientific and technological development, and shall be the province of all mankind. Particular account should be taken of the needs of developing countries.

Furthermore the Resolution provides that States are free to determine all aspects of their participation in international cooperation in the exploration and use of outer space on an equitable and mutually acceptable basis. Contractual terms in such cooperative ventures should be fair and reasonable and they should be in full compliance with the legitimate rights and interests of the parties concerned, as, for example, with intellectual property rights. All States, particularly those with relevant space capabilities and with programmes for the exploration and use of outer space, should contribute to promoting and fostering international cooperation on an equitable and mutually acceptable basis. In this context, particular attention should be given to the benefit and the interests of developing countries and countries with incipient space programmes stemming from such international cooperation conducted with countries with more advanced space capabilities.

Contained in the Resolution 51/122 is also the principle that international cooperation should be conducted in the modes that are considered most effective and appropriate by the countries concerned, including, inter alia, governmental and non-governmental; commercial and non-commercial; global, multilateral, regional or bilateral; and international cooperation among countries in all levels of development. International cooperation, while taking into particular account the needs of developing countries, should aim, inter alia, at the following goals, considering their need for technical assistance and rational and efficient allocation of financial and technical resources:

---

[3] Declaration on International Cooperation in the Exploration and Use of Outer Space for the Benefit and in the Interest of All States, Taking into Particular Account the Needs of Developing Countries.

(a) Promoting the development of space science and technology and of its
    applications;
(b) Fostering the development of relevant and appropriate space capabilities in
    interested States;
(c) Facilitating the exchange of expertise and technology among States on a
    mutually acceptable basis.

National and international agencies, research institutions, organizations for
development aid, and developed and developing countries alike should consider the
appropriate use of space applications and the potential of international cooperation
for reaching their development goals. The United Nations Committee on the
Peaceful Uses of Outer Space (UNCOPUOS)[4] should be strengthened in its role,
among others, as a forum for the exchange of information on national and international activities in the field of international cooperation in the exploration and use
of outer space. All States should be encouraged to contribute to the United Nations
Programme on Space Applications and to other initiatives in the field of international cooperation in accordance with their space capabilities and their participation
in the exploration and use of outer space.

5.2 ICAO and UNCOPUOS

ICAO's Secretary General, in his letter of invitation to Commander Hadfield,
concludes by saying, "In closing, let me quote from the preamble to the Convention
on International Civil Aviation that created ICAO seventy years ago, on 7
December 1944. The drafters of the Convention understood the potential of aviation
to create and preserve friendship and understanding among the nations and peoples
of the world", underscoring that it is desirable to promote cooperation between

---

[4] In 1958 The United Nations established the United Nations Committee on the Peaceful Uses of
Outer Space (UNCOPUOS), an ad hoc committee in order to consider the activities and resources
of the United Nations, the specialized agencies and other international bodies relating to the
peaceful uses of outer space international cooperation and programmes in the field that could
appropriately be undertaken under United Nations auspices;
   organizational arrangements to facilitate international cooperation in the field within the
framework of the United Nations; and legal problems which might arise in programmes to explore
outer space. In 1959, the General Assembly established the Committee as a permanent body and
reaffirmed its mandate in Resolution 1472 (XIV). In 1961, the General Assembly, considering that
the United Nations should provide a focal point international cooperation in the peaceful exploration and use of outer space, requested the Committee, in cooperation with the Secretary-General
and making full use of the functions and resources of the Secretariat; to maintain close contact with
governmental and non-governmental organizations concerned with outer space matters; to provide
for the exchange of such information relating to outer space activities as Governments may supply
on a voluntary basis, supplementing, but not duplicating, existing technical and scientific
exchanges; and to assist in the study of measures for the promotion of international cooperation in
outer space activities.

nations and peoples upon which the peace of the world depends. The same can be said of space travel and this is what we would like to explore with you.

The overall thrust of the Secretary General's vision in this process is seemingly in the intent to pursue deliberations and discussions along the aviation theme of friendship and understanding, which he correctly attributes to space transportation as well. In this regard, ICAO's close association with UNCOPUOS would be significant. The subject of ICAO's involvement in outer space activities was addressed by the author in 2004 in a publication in the *Journal of Space Law*.[5] At that point, ICAO had not even considered such an involvement. Five years after this publication, the Legal Subcommittee of UNCOPUOS , at its forty-eighth session, in 2009, continued to address the item "Matters relating to the definition and delimitation of outer space and the character and utilization of the geostationary orbit, including consideration of ways and means to ensure the rational and equitable use of the geostationary orbit without prejudice to the role of the International Telecommunication Union". The Committee then requested ICAO to make, at the forty-ninth session of the Subcommittee, in 2010, a comprehensive presentation on current and foreseeable civil aviation operation, with particular emphasis on the upper limit of those operations. It must be noted that UNCOPUOS did not request ICAO to present its views on outer space activities or space transport but merely sought ICAO's views on "current and foreseeable civil aviation operation, with particular emphasis on the upper limit of those operations".

ICAO responded to this request in 2010 by stating that it is aware of the fact that commercial suborbital operations are being planned by various entities and expect that such operations would affect international civil aviation in some manner, without specifying how international civil aviation would be affected.

UNCOPUOS encourages national regulation of space activities and may have its own view of a global organization being involved in one form or another in the field of regulating commercial space activities. At its Fifty Seventh Session in June 14 2014 the Committee agreed that a panel discussion addressing sustainable development within the context of the post-2015 development agenda should be organized by the Office for Outer Space Affairs and be held in the Fourth Committee of the General Assembly during the sixty-ninth session of the Assembly, when it considers the item "International cooperation in the peaceful uses of outer space", and that the specific topic of the panel discussion should be determined by the Office. Given the link the proposed panel discussion would have with ICAO symposium, it would be interesting to see the participation of UNCOPUOS in the ICAO discussions.

The United Nations General Assembly, adopted in 2005 Resolution 59/115 which recommended that that States conducting space activities, in fulfilling their international obligations under the United Nations treaties on outer space, in particular the Treaty on Principles Governing the Activities of States in the Exploration and Use of Outer Space, including the Moon and Other Celestial Bodies, the

---

[5] Abeyratne 2004, pp. 185–205.

Convention on International Liability for Damage Caused by Space Objects1 and the Convention on Registration of Objects Launched into Outer Space, as well as other relevant international agreements, consider enacting and implementing national laws authorizing and providing for continuing supervision of the activities in outer space of non-governmental entities under their jurisdiction. The Resolution also recommended that States consider the conclusion of agreements in accordance with the Liability Convention with respect to joint launches or cooperation programme and that UNCOPUOS invite Member States to submit information on a voluntary basis on their current practices regarding on-orbit transfer of ownership of space objects. Another recommendation was that States consider, on the basis of that information, the possibility of harmonizing such practices as appropriate with a view to increasing the consistency of national space legislation with international law. UNCOPUOS was requested to continue to provide States, at their request, with relevant information and assistance in developing national space laws based on the relevant treaties in making full use of the functions and resources of the Secretariat.

References

Abeyratne R (2004) ICAO's involvement in outer space affairs—a need for closer scrutiny. J Space Law 30(2):185–202

Dembling PG, Arons DM (1968) The treaty on rescue and return of astronauts and space objects. William Mary Law Rev 9:630–663 at 643

Chapter 6
Legal Legitimacy of ICAO and Direction to Be Taken

6.1 Legal Legitimacy

6.1.1 Applicable Principles

International organizations can generally only work on the basis of legal powers that
are attributed to them. Presumably, these powers emanate from the sovereign States
that form the membership of such organizations.[1] Therefore, the logical conclusion
is that if international organizations were to act beyond the powers accorded to them,
they would be presumed to act *ultra vires*.[2] A seminal judicial decision relating to the
powers of international organizations was handed down by the Permanent Court of
International Justice in 1922 in a case[3] relating to the issue as to whether the
International Labour Organization (set up to regulate international labour relations)
was competent to regulate labour relations in the agricultural sector. The court
proceeded on the basis that the competence of an international organization with
regard to a particular function lay in the treaty provisions applicable to the functions
of that organization and that the determination of such competence would be based
on interpretation. In this instance the Court was of the view that, in its interpretation
of the ILO treaty, the organization had the power to extend its scope of functions to
the agricultural sector. Based on the wide scope of universality of ICAO recognized
by the Chicago Conference, one could conclude without doubt that the principle of
the ILO case could be applied to ICAO when it extends its functions to areas that are
strictly non aviation, such as the development of technical specifications for the
Machine Readable Travel Document (MRTD) which, although is not directly related
to the techniques of air navigation and the economics of air transport is nonetheless

---

[1] See de Witte 1998, pp. 277–304.
[2] See Klabbers 2002, p. 60.
[3] *Competence of the ILO to regulate the Conditions of Labour of Persons Employed in Agriculture*, Advisory Opinion [1922] Publ. PCIJ Series B, nos. 2&3.

© The Author(s) 2015
R. Abeyratne, *Regulation of Commercial Space Transport*,
SpringerBriefs in Law, DOI 10.1007/978-3-319-12925-9_6

related to aviation and therefore impact the development and well being of the aviation industry. A good example of this principle of implied extension is ICAO's work on machine readable travel documents and the administration of the Public Key Directory.[4] However, the principle of implied extension should be carefully applied, along the fundamental principle enunciated by Judge Green Hackworth in the 1949 *Reparation for Injuries Case*[5]—that powers not expressed cannot freely be implied and that implied powers flow from a grant of express powers, and are limited to those that are necessary to the exercise of powers expressly granted.[6]

The universal solidarity of ICAO Contracting States that was recognized from the outset at the Chicago Conference brings to bear the need for States to be united in recognizing the effect of ICAO policy and decisions. This principle was given legal legitimacy in the *ERTA* decision[7] handed down by the Court of Justice of the European Community in 1971. The court held that the competence of the European Community to conclude an agreement on road transport could not be impugned since the member States had recognized Community solidarity and that the Treaty of Rome which governed the Community admitted of a common policy on road transport which the Community regulated.

It should be noted that ICAO does not only derive implied authority from its Contracting States based on universality but it also has attribution from States to exercise certain powers. The doctrine of attribution of powers comes directly from the will of the founders, and in ICAO's case, powers were attributed to ICAO when it was established as an international technical organization and a permanent civil aviation agency to administer the provisions of the Chicago Convention. In addition, ICAO could lay claims to what are now called "inherent powers" which give ICAO power to perform all acts that the Organization needs to perform to attain its aims not due to any specific source of organizational power but simply because ICAO inheres in organizationhood. Therefore, as long as acts are not prohibited in ICAO's constituent document (the Chicago Convention), they must be considered legally valid.[8]

Over the past two decades the inherent powers doctrine has been attributed to the United Nations Organization and its specialized agencies on the basis that such organizations could be stultified if they were to be bogged down in a quagmire of interpretation and judicial determination in the exercise of their duties. The advantages of the inherent powers doctrine is twofold. Firstly, inherent powers are

---

[4] The public key directory is designed and proposed to be used by customs and immigration authorities who check biometric details in an electronic passport, and is based on cryptography. It is already a viable tool being actively considered by the aviation community as a fail-safe method for ensuring the accuracy and integrity of passport information. For a detailed discussion on the public key directory see. Abeyratne 2005, pp. 255–268.

[5] *Reparation for Injuries Suffered in the Service of the United Nations*, advisory opinion, [1949] ICJ Reports 174.

[6] *Id.* at p. 198.

[7] Case 22/70, *Commission* v. *Council* (European Road Transport Agreement) [1971] ECR 273.

[8] See Seyersted 1963, p. 28.

functional and help the organization concerned to reach its aims without being tied by legal niceties. Secondly, it relieves the organization of legal controls that might otherwise effectively preclude that organization from achieving its aims and objectives. The ability to exercise its inherent powers has enabled ICAO to address issues on aviation insurance and establish an insurance mechanism; perform mandatory audits on States in the fields of aviation safety and security; and establish a funding mechanism to finance aviation safety projects, all of which are not provided for in the Chicago Convention but are not expressly prohibited.

With regard to the conferral of powers by States to ICAO, States have followed the classic approach of doing so through an international treaty. However, neither is there explicit mention of such a conferral on ICAO in the Chicago Convention nor is there any description of ICAO's powers, except for an exposition of ICAO's aims and objectives. The Council of ICAO is designated both mandatory and permissive "functions", although, as already discussed, the Council could impose certain measures when provisions of the Convention are not followed. Therefore States have not followed the usual style of conferral of powers in the case of ICAO, which, along the lines of the decision of the International Court of Justice in the 1996 *WHO Advisory Opinion* case[9] was that the powers conferred on international organizations are normally the subject of express statement in their constituent instruments.[10] This notwithstanding, it cannot be disputed that ICAO Contracting States have conferred certain powers on ICAO to perform its functions independently. For example, ICAO is a legal entity having the power to enter into legal agreements with legal entities including other international organizations with regard to the performance of its functions.

Conversely, an international organization must accept conferred powers on the basis of Article 34 of the Vienna Convention on the Law of Treaties which stipulates that a treaty does not create rights or obligations of a third State without its consent. This principle can be applied *mutatis mutandis* to an international organization such as ICAO. The conferral of powers on an international organization does not *ipso facto* curtail the powers of a State to act outside the purview of that organization unless a State has willingly limited its powers in that respect. This principle was recognized in the *Lotus* Case[11] where the Provisional International Court of Justice held that a State can exercise powers on a unilateral basis even while the conferral to the Organization remains in force. The Court held that restrictions upon the independence of States cannot be presumed.[12]

ICAO's conferred powers enable the Organization to adopt binding regulations by majority decision (which is usually unnecessary as most of ICAO policy is adopted through consensus). However, States could opt out of these policies or make reservations thereto, usually before such policy enters into force. This is

---

[9] *Legality of the Threat or Use of Nuclear Weapons, Advisory Opinion*, ICJ Reports, 1996, p. 64.

[10] *Id.* p. 79.

[11] *PCIJ Reports* Series A, No. 10, p. 4.

[12] *Id.* p. 18.

because States have delegated power to ICAO to make decisions on the basis that they accept such decisions on the international plane. In such cases States could contract out and enter into binding agreements outside the purview of ICAO even on subjects on which ICAO has adopted policy. The only exception to this rule lies in the adoption of Standards in Annex 2 to the Chicago Convention on Rules of the Air, in particular navigation over the high seas and other over flight areas where freedom of flight prevails which all Contracting States are bound to follow in order to maintain global safety.

Given ICAO's nature as a self standing legal entity, the Organization would be responsible for its internationally wrongful acts.[13] As to the issue whether a State which has delegated powers to ICAO would be responsible for the wrongful acts of the Organization, a State is not bound by the Organization's exercise of delegated powers[14] and therefore it cannot be necessarily assumed that such acts would be attributable to the States unless such acts were the effect of the State's own acts or omissions. Article 1 of the *Articles of Responsibility* of the International Law Commission (ILC) expressly stipulates that every internationally wrongful act entails the international responsibility of a State.[15] The State cannot escape responsibility by seeking refuge behind the non-binding decision of an Organization in the case of delegation of powers. This is also the case where a State aids and abets an Organization to perform an internationally wrongful act.[16]

ICAO is an Organization established by the Chicago Convention and is composed of an Assembly, a Council and such other bodies as may be necessary. The Assembly and the Council are composed of Contracting States. The Assembly, composed of 189 Contracting States, has delegated its daily functions to the Council of 36 Member States which largely forms the decision making body of ICAO. Therefore it would not be incorrect to assume that any resolution adopted or decision taken by the ICAO Council can be imputed to ICAO's Contracting States which have delegated powers on the Council. However, States retain the powers to act unilaterally and they are not bound to comply with obligations flowing from the Organization's exercise of conferred powers. States which have delegated powers on ICAO have the legal right under public international law to take measures against a particular exercise by ICAO of conferred powers which is considered to

---

[13] G. Gaja, *First Report on Responsibility of International Organizations*, A/CN.4/532, ILC 56th Session, p. 19 para. 35.

[14] In the case of agency, where an organization acts as the agent of States the answer would be in the affirmative, based on the fact that the control exercised by States over the Organization meant that *prima facie*, the acts of the Organization were attributable to the States. *Id*. Chapter 4, Section III(3)(ii).

[15] See Crawford 2002, p. 77.

[16] Article 16 of the ILC's Articles of Responsibility provides that a State which aids or abets another State in the the commission of an internationally wrongful act by the latter is responsible for doing so if that State does so with the knowledge of the circumstances of such act and if the act would be internationally wrongful if committed by that State. This principle can be applied mutatis mutandis to an organization such as ICAO.

be *detournement de pouvoir*, *ultra vires* or an internationally wrongful act with which the objecting States do not wish to be associated. A State could also distance itself from the State practice of other Contracting States within the Council if such activity is calculated to form customary international law that could in turn bind the objecting State if it does not persist in its objections.[17]

As discussed earlier in this book, a significant issue in the determination of ICAO's effectiveness as an international organization is the overriding principle of universality and global participation of all its 191 Contracting States in the implementation of ICAO policy. This principle, which has its genesis in the Chicago Conference of 1944, has flowed on gaining express recognition of legal scholars. This is what makes ICAO unique as a specialized agency of the United Nations and establishes without any doubt that ICAO is not just a tool of cooperation among states.

ICAO is established by the Chicago Convention, but there is no explicit description of the Organization in the Convention except for a detailed expose of its aims and objectives in Article 44. This could well be due to the fact that the notion of an international organization to handle international civil aviation was established during the Chicago Conference of 1944 before the establishment of the United Nations, of which ICAO later became a specialized agency. Therefore, one has to go back to the discussions and proceedings of the Conference in order to determine the intention of the parties which led to establishing the meaning and purpose of ICAO.

Although at the Conference several proposals were placed on the table by the United States,[18] United Kingdom,[19] Canada[20] and Australia and New Zealand jointly,[21] consensus was finally reached on establishing an international technical organization—a permanent civil aviation agency to administer the provisions of the Chicago The distinguishing feature of ICAO was that its founding fathers were firmly of the view that an international technical organization such as ICAO must have universality and that the degree of urgency of universality was greater in the case of ICAO than, for example, the International Monetary Fund or the Food and

---

[17] See Sarooshi 2005, p. 110.

[18] The United States called for an international aviation authority with powers limited to the technical and consultative level in the economic field. See *Proceedings of the International Civil Aviation Conference*, Chicago, Illinois, November 1–December 7, 1944, Department of State Publication 2820, Washington, 1948, Vol. II, pp. 1317–1319.

[19] The United Kingdom aimed to give more discretionary authority in allocating routes, fixing rates and determining frequencies. See *Proceedings of the International Civil Aviation Conference*, *id.* at p. 63.

[20] The Canadian proposal envisioned the establishment of a democratic world air assembly in which each nation would have two votes. The Canadian proposal was for international regulation and not for international ownership or international operation. See *Proceedings of the International Civil Aviation Conference, Supra* note 281, at p. 67.

[21] The joint Australia–New Zealand proposal was for the establishment of international control and operation of all international airways, including international ownership. See *Proceedings of the International Civil Aviation Conference, id.*, at pp. 77–80.

Agriculture Organization.[22] In support of this approach, Schenkman offers the practical illustration that if a non-member of ICAO were to operate an international air service that did not adhere to established ICAO rules of the air, the safety of operations of air services of all member States would be jeopardized.[23] From an economic perspective, the same commentator quotes the instance of a non member State closing its air space to other nations, thus effectively precluding economic progress of the air transport industry. These examples illustrate the distinction earlier made between ICAO and other international organizations such as FAO and IMF where a non participating State might not do as harm to other States in its actions in the food and agriculture or monetary areas as much as it would in the field of air navigation and transport. Therefore, universal acceptance of ICAO by its member States was considered a *sine qua non* by the founding fathers of the Organization. Over the years this principle has grown in strength particularly in view of the exponential growth of air transport and the vast technological advances made in the field of air navigation and air transport.

It is worthy of note that the ICAO Assembly, at its First Session held in Montreal from 6 to 27 May 1947 resolved that:

> Universal membership of in the International Civil Aviation Organization is desirable to achieve its maximum usefulness in promoting safety in the air and the efficient and orderly development of air transport.[24]

Therefore, it is incontrovertible that universal participation in ICAO by the contracting States is indispensable if ICAO were to effectively implement the provisions of the Chicago Convention. Sixty years of symbiotic existence have shown that States need ICAO no less than ICAO needs their membership.

ICAO is primarily governed by international law, being recognized by the United Nations Charter as a specialized agency of the United Nations. It is also governed by two major agreements, one between the United Nations and ICAO and the other between the Government of Canada and ICAO. The Headquarters Agreement between ICAO and Canada,[25] in Article 2, explicitly provides that ICAO shall possess juridical personality and shall have the legal capacities of a body corporate including the capacity to contract; to acquire and dispose of movable and immovable property; and to institute legal proceedings in Canada Article 3 of the Agreement stipulates that ICAO, its property and its assets,

---

[22] See Schenkman 1955a, p. 125.

[23] *Ibid.*

[24] Resolution A1-9, ICAO Doc 7375-C/852, 1947 pp. 242–243. The same philosophy of universality was extended to the United Nations in general at the first session of the United Nations—that the United Nations looked forward to the time when universal membership in the system of the international organization established by the Charter of the United Nations would be possible and actual. Schenkman, *supra* note 285 at p. 87.

[25] Headquarters Agreement Between the International Civil Aviation Organization and the Government of Canada, ICAO Doc 9591. For further information on the agreement see Michael Milde, New Headquarters Agreement Between ICAO and Canada, (1992) *Annals Air and Sp. L.* Part II, 305–322.

wherever located and by whomsoever held, shall enjoy the same immunity from suit and every form of judicial processes as is enjoyed in foreign States. (Assets include funds administered by the Organization in furtherance of its constitutional functions.)

The headquarters premises of ICAO is inviolable and is given the same protection by the Government of Canada as is given to diplomatic missions in Canada.[26] The organization, its assets, income and property, owned or occupied in Canada are exempt from taxes[27] as well as goods purchased under appropriate certificates from manufacturers or wholesalers who are licensed under the Excise Act.[28]

A seminal judicial decision relating to the powers of international organizations was handed down by the Permanent Court of International Justice in 1922 in a case[29] relating to the issue as to whether the International Labour Organization (set up to regulate international labour relations) was competent to regulate labour relations in the agricultural sector. The court proceeded on the basis that the competence of an international organization with regard to a particular function lay in the treaty provisions applicable to the functions of that organization and that the determination of such competence would be based on interpretation. However, the principle of implied extension should be carefully applied, along the fundamental principle enunciated by Judge Green Hackworth in the 1949 *Reparation for Injuries Case*[30]—that powers not expressed cannot freely be implied and that implied powers flow from a grant of express powers, and are limited to those that are necessary to the exercise of powers expressly granted.[31]

States retain the powers to act unilaterally and they are not bound to comply with obligations flowing from the Organization's exercise of conferred powers. States which have delegated powers on ICAO have the legal right under public international law to take measures against a particular exercise by ICAO of conferred powers which is considered to be *detournement de pouvoir, ultra vires* or an internationally wrongful act with which the objecting States do not wish to be associated. A State could also distance itself from the State practice of other Contracting States within the Council if such activity is calculated to form customary international law that could in turn bind the objecting State if it does not persist in its objections.[32]

---

[26] Headquarters Agreement id, Article 4(1) and (2).

[27] *Id.* Article 6.

[28] *Id.* Article 7.

[29] *Competence of the ILO to regulate the Conditions of Labour of Persons Employed in Agriculture*, Advisory Opinion [1922] Publ. PCIJ Series B, nos. 2&3.

[30] *Reparation for Injuries Suffered in the Service of the United Nations*, advisory opinion, [1949] ICJ Reports 174.

[31] *Id.* at p. 198.

[32] See Sarooshi 2005, p. 110.

ICAO's identity before courts having national jurisdiction would strictly be restricted to the nature of the organization and the type of work it carries out. Any special privilege accorded to ICAO by agreement or treaty would therefore be applicable only in relation to ICAO's scope of work.[33] Conceptually, it has been argued that in an instance of national litigation involving an international organization, courts would, in the event the litigious issue pertains to the work of that organization, apply the "functional theory" in an *acta jure gestionis* (commercial act), which means that the organization concerned will not be viewed as having special immunities or privileges. In the 1953 case of *Re International Bank for Reconstruction and Development and International Monetary Fund* v. *All America Cables and Radio Inc., and other cable companies*[34] the US Federal Communications Commission was confronted with the argument of the plaintiffs—the World Bank and the IMF—that the purpose of granting privileges and immunities to organizations located in the jurisdiction of a State where national law applied to contracts is to protect such organizations from unfair and undue interference including excessively high rates. The defendant (radio and cable) companies argued that there was no evidence or reason to allow the banks lower-than-commercial rates. The rationale that can be drawn from this case is that the purpose of immunity will be destitute of effect if courts were asked to determine the legality of an organization's work if such inquiry were to obstruct the work of that organization.

A question arises as to what extent or within what parameters must a court apply the principle of functional immunity to commercial acts of an international organization. Courts have veered from one extreme, coming close to recognizing absolute immunity as in the case of *Broadbent* v. *Organization of American States*[35] to linking key activities of an organization, such as its interpretation and translation services to *acta jure imperii* (sovereign act) on the basis that language services were integral to the main functions of an organization.[36]

It is curious that, seven decades after the establishment of ICAO, some still refer to its powers and functions.[37] There are some others who allude to ICAO's mandate. The fact is that ICAO has only aims and objectives, recognized by the Chicago Convention which established the Organization. Broadly, those aims and objectives are to develop the principles and techniques of international air navigation and to foster the planning and development of international air transport. In effect, this bifurcation implicitly reflects the agreement of the international community of States which signed the Chicago Convention that ICAO could adopt Standards in the technical fields of air navigation and could only offer guidelines in

---

[33] In *United States* v. *Malekh* et al., 32 ILR 308–334 (1960) where the defendant, a United Nations employee, was charged with espionage, the US District Court for the Southern District of New York held that neither the defendant nor his employer should have any claims to immunity as espionage was not a part of the functions of the United Nations.

[34] 22 *ILR* 705-712.

[35] 63 *ILR* 162-163: US District Court for the District of Columbia, 28 March 1978.

[36] *Iran-United States Claims Tribunal* v. *A.S.*, 94 ILR 321–330.

[37] See MacKenzie 2008, Preface at p. 1.

the economic field. This exclusive right in the technical field initially bestowed on ICAO the authority to set standards for equipment and procedures on international air routes in the first years (1947–1949) of ICAO.[38] This key task of standardizing technical specifications in air navigation gave rise to the realization that, apart from States that were most advanced in technology and could implement ICAO's standards, there were numerous other States who could not implement the Standards, however, willing they would be, due to the lack of resources and know-how.[39] This gave rise to offers of help by numerous member States of ICAO, and the technical assistance limb of ICAO was born.

One has, however, to note that the position of ICAO as an incipient regulator was quite different from the one it is placed in now. In 1949 ICAO was feeling the pulse of the world of civil aviation and formulating regulations. This was the year in which ICAO embarked upon a comprehensive study of the ground facilities operated by governments and the services that they provided.[40] Traffic volumes were only beginning to pick up and passengers and cargo between North America and Europe were being carried by eleven carriers. Elsewhere, in South America direct east-west services went into operation for the first time.[41] ICAO has a busy work schedule both in the technical and economic fields. The current and future role of ICAO does not therefore have to be reinvented. However, for the past 60 years or so, ICAO has been active in its standardization role, which has been blended in recent years with a burgeoning implementation role that is gradually blurring the former. In a world that is becoming largely globalized and regionalized, ICAO has vastly to focus on not so much what it does but how it does its work. In this context, ICAO has a dual role to play. The first is to act as a global forum for aviation, which is primarily the role expected of ICAO by the developed nations which are largely self reliant in regulatory matters. However, they need ICAO to set global standards that could apply to all ICAO's 191 member States. On the other hand, ICAO has to be both a global forum and a mentor to the developing world which expects ICAO to assist and guide them.

In view of its clear empowerment by the States ICAO may have to rely on extending its aims and objectives to cover the development of principles and techniques of commercial space travel. Unless otherwise agreed by ICAO's member States, ICAO's existing aims and objectives, as recognized by Article 44 of the Chicago Convention would extend to insuring the safe and orderly growth of commercial space transport throughout the world; encouraging the arts of spacecraft design and operation for peaceful purposes; encouraging the development of

---

[38] Max Hymans, President of the Second ICAO Assembly observed in 1948 that the standardization of equipment and procedures had progressed well and that ICAO had an incontestable authority in this respect. See Max Hymens, Results of a Meeting, *Interavia* 3, 8(August 1948) at p. 422.

[39] See. ICAO Doc 6968, A4-P/1 Report of the Council to the Assembly on the Activities of the Organization in 1949, 23 March 1950.

[40] *Id.* A4-P/1 23/3/50 at p. 1.

[41] *Id.* 2.

airways, spaceports, and air navigation facilities insofar as they relate to commercial space transport; meeting the needs of the peoples of the world for safe, regular, efficient and commercial space transport; preventing economic waste caused by unreasonable competition; Insuring that the rights of Contracting States are fully respected and that every Contracting State has a fair opportunity to operate commercial space transport and spacecraft airlines; avoiding discrimination between Contracting States; promoting safety of space flight in international air navigation; and promoting generally the development of all aspects of aerospace engineering and technology.

The two main characteristics of ICAO are: that it is created by States, more specifically, as States themselves are abstractions, by duly authorized representatives of States; and that they are created by treaty, which is a written agreement signed by the States' Parties to it and governed by international law. States can only act by and through their agents. Different government departments or Instrumentalities of State bear responsibility for different international organizations. In the case of ICAO, the most likely government department that would be responsible for the Organization within a State would be the Ministry or departments of transport or aviation as the case may be. The third characteristic that distinguishes an international organization as a "club" of States without just being the spokesperson or mouthpiece of those States is that it is expected to have a "will" of its own. ICAO's independent will, recognized by the Government of Canada for purposes of its activities within the country is encapsulated in a provision in the *Headquarters Agreement between ICAO and the Government of Canada* which states that ICAO has an identity of its own, capable of entering into contracts. This having been said, ICAO is by no means sovereign in its own right, although courts have on occasion referred to sovereign rights of an organization merely to seek a compromise between absolute acceptance of parity between a State and an organization and absolute refusal of an international organization's ability to perform *acta jure imperii* (governmental acts).

Within the parameters of its empowerment which ICAO attenuates from its member States ICAO should, in both air and space matters make a philosophical adjustment with a view to ensuring that it keep abreast of the new world order where States are increasingly being disaggregated into components which act in global networks, linking the world together in a manner that enables global trends to permeate the local environment. In other words, ICAO should facilitate interaction between States and their components that interact in matters of civil aviation and commercial space transport. For example, in many member States, aviation has numerous players in different areas such as customs and immigration, medical and quarantine, tourism, police, airports and air navigation service providers. In most instances these players do not act in accord, thus resulting in disharmony in the ultimate delivery of an efficient air transport product. This could well be the case also in space transport. ICAO's Mission and Vision Statement exhorts ICAO to do just what is needed—to in acting as the global forum in the key areas of concern to international civil aviation through cooperation between its member States.

While promoting fluid dialogue and cooperation among its member States, ICAO should take the initiative to assist States both in technical and economic issues. This assistance is not confined to providing technical assistance through projects administered by the Technical Cooperation Bureau but should also extend to providing guidance, mainly to States which still look up to ICAO as the global forum of aviation experts.

ICAO could also give consideration to its role in technical cooperation in commercial space activities. ICAO could cooperate with States and other relevant entities to provide technical assistance taking into account the priorities of States and ICAO policy in implementing the Strategic Objectives[42] of ICAO. This cooperation will be forged primarily through projects, particularly where such projects are necessary for the provision of vital air transport infrastructure and/or the economic development of a State. In implementing this Policy, ICAO will optimally use its resources both at Headquarters and its Regional Offices and apply the principles enunciated in the relevant ICAO Assembly Resolutions, guidance and policy.

Overall responsibility for the implementation and continued evolution of this Policy devolves upon the Secretary General of ICAO, assisted by the Director, Technical Cooperation Bureau. As necessary and where relevant, this Policy will be incorporated into the ICAO workplace through the Organization's Business Plan.

6.1.2 Technical Assistance

It is an interesting fact that the Chicago Convention, from which ICAO derives its legal legitimacy, does not contain a single reference to technical assistance to be rendered by ICAO. Yet the Technical Cooperation Programme (TCP)[43] of ICAO which is executed by the Technical Cooperation Bureau (TCB) that was established in 1952, has maintained a sustained record of technical assistance provided to States over a span of 60 years. TCP is the major operational tool for reinforcing the Organization's technical cooperation mission objectives, including enhancing the capacity of developing countries to implement ICAO Standards and Recommended

---

[42] ICAO's Strategic Objectives pertain to safety—enhancement of global civil aviation safety; air navigation capacity and efficiency—increasing capacity and improving efficiency of the global civil aviation system; security and facilitation—enhancement of global civil aviation security and facilitation; economic development of air transport—fostering the development of a sound and economically-viable civil aviation system; and environmental protection—minimizing the adverse environmental effects of civil aviation activities.

[43] ICAO's Technical Co-operation Programme provides advice and assistance in the development and implementation of projects across the full spectrum of civil aviation aimed at the safety, security, environmental protection and sustainable development of national and international civil aviation. The Programme is conducted under the broad policy guidance of the ICAO Assembly and of the Council. Subject to general guidance by the Secretary General, the Technical Co-operation Programme is executed by the Technical Co-operation Bureau (TCB).

Practices (SARPs). Its continuing importance has been reaffirmed by the Assembly in several resolutions, *inter alia* in the Consolidated Statement of ICAO Policies on Technical Co-operation (Resolution A36-17), which stipulates that the Technical Co-operation Programme is a permanent priority activity of ICAO that complements the role of the Regular Programme in providing support to States in the effective implementation of SARPs and Air Navigation Plans (ANPs) as well as in the development of their civil aviation administration infrastructure and human resources; and is furthermore one of the main instruments of ICAO to assist States in remedying the safety and security deficiencies identified through ICAO's audit programmes.

6.1.3 Dispute Settlement

ICAO could also give detailed consideration to its dispute settlement in commercial space transportation activities. It should not be forgotten that another positive feature of ICAO, which lends itself well to the Organization's possible involvement in outer space activities, is its proven competence in mediation during the settlement of disputes. A preeminent feature of the ICAO Council is its indomitable resolve to address its deliberations to purely technical issues pertaining to any dispute, while stringently avoiding political issues and pitfalls. This is certainly true of all disputes brought before the Council, where the Council restricted its scope to technical issues as applicable to the principles embodied in the Chicago Convention. At present The Council has therefore the power under the Chicago Convention to adjudicate disputes between the member States of ICAO on matters pertaining to international civil aviation. As already mentioned, the Council is a permanent body responsible to the ICAO Assembly and is composed of 36 Contracting States elected by the Assembly. It has its genesis in the Interim Council of the Provisional International Civil Aviation Organization (PICAO).[44] PICAO occupied such legal capacity as may have been necessary for the performance of its functions and was recognised as having full juridical personality wherever compatible with the Constitution and the laws of the State concerned.[45] The definitive word "juridical" attributed to PICAO a mere judicial function, unequivocally stipulating that the organization and its component bodies, such as the Interim Council were obligated to remain within the legal parameters allocated to them by the Interim Agreement[46] and that PICAO was of a purely technical and advisory nature. A legislative or

---

[44] Hereafter referred to as PICAO. See Interim Agreement on International Civil Aviation, opened for signature at Chicago, December 7 1944, Article 3. Also in Hudson, *International Legislation*, Vol. 1X (1942–1945, New York) at p. 159.

[45] *Id.* Article 1 Section 4. It is interesting to note that PICAO was established as a provisional organization of a technical and advisory nature for the purpose of collaboration in the field international civil aviation. *Vide* Article 1 Section 1.

[46] *Op. cit.*

quasi-legislative function could not therefore be imputed to the Interim Council of PICAO. It could mostly study, interpret and advise on standards and procedures[47] and make recommendations with respect to technical matters through the Committee on Air Navigation.[48] The International Civil Aviation Organization (ICAO) which saw the light of day on April 4, 1947 derived the fundamental postulates of its technical and administrative structure from its progenitor—PICAO—and it would seem reasonable to attribute a certain affinity *ipso facto* between the two organizations and hence, their Councils. One of the Council's functions is to consider any matter relating to the Convention which any Contracting State refers to it.[49] Since one of the distinctive features of the ICAO Council is its ability to make rules for international civil aviation, it follows incontrovertibly that the Council's dispute resolution powers are compelling.

The ICAO Council has played a signal role in dispute resolution in the nineties up to date. Over the past two decades. One of the best examples of ICAO's role in the international community was seen in The Iran air Incident—IR 655 (Iran, United States 1998). This concerned the shooting down of an Iran Air Airbus A300 (IR655) carrying commercial passengers on a scheduled flight from Bandar-Abbas (Iran) to Dubai. The aircraft was brought down by the *U.S.S. Vincennes* over the Persian Gulf, resulting in the death of all 290 persons on board the aircraft.

One of the emergent features of the ICAO Council which became clear at its deliberations was the Council's resolve to address its deliberations to purely technical issues pertaining to the incident, while stringently avoiding political issues and diplomatic pitfalls. This is certainly true of all incidents discussed above, where the Council restricted its scope to technical issues as applicable to the principles embodied in the Chicago Convention

However, there seems to be an unfortunate dichotomy in terminology in the Convention since on the one hand, Article 54(n) makes it mandatory that the Council shall merely consider any matter relating to the Convention which any Contracting State refers to it, while on the other, Article 84 categorically states that any disagreement between two or more States relating to the interpretation or application of the Convention and its Annexes, that cannot be settled by negotiation shall…be decided by the Council. The difficulty arises on a strict interpretation of Article 54(n) where even a disagreement between two States as envisaged under Article 84 could well be considered as 'any matter' under Article 54(n). In such an instance, the Council could well be faced with the dilemma of choosing between the two provisions. It would not be incorrect for the Council to merely consider a matter placed before it, although a decision is requested by the applicant State, since, Article 54(n) is perceived to be comprehensive as the operative and controlling provision that lays

---

[47] Interim Agreement, *supra* note 44, Section 6.4.b(1).

[48] *Id.* Section 6.4.b(6). Also, T. Buergenthal, infra note 381
    at 4, where the author states that PICAO's functions were merely advisory, which precludes any imputation of legislative or quasi-legislative character to its Interim Council.

[49] Chicago Convention, supra note 1 Article 54.

down mandatory functions of the Council. It is indeed unfortunate that these two provisions obfuscate the issue which otherwise would have given a clear picture of the decision making powers of the Council. A further thread in the fabric of adjudicatory powers of the Council is found in Article 14 of the Rules of Settlement promulgated by the Council in 1957[50] which allows the Council to request the parties in dispute to engage in direct negotiations at any time.[51] This emphasis on conciliation has prompted the view that the Council, under Article 84 would favour the settling of disputes rather than adjudicating them.[52] This view seems compatible with the proposition that the consideration of a matter under Article 54(n) would be a more attractive approach *in limine* in a matter of dispute between two States.

Milde noted in 1979:

> The Council of ICAO cannot be considered a suitable body for adjudication in the proper sense of the word - i.e. settlement of disputes by judges and solely on the basis of respect for law. The Council is composed of States (not independent individuals) and its decisions would always be based on policy and equity considerations rather than on pure legal grounds...truly legal disputes...can be settled only by a true judicial body which can bring into the procedure full judicial detachment, independence and expertise. The underemployed ICJ is the most suitable body for such types of disputes.[53]

The perceived inadequacies of the ICAO Council in being ethically unsuitable to decide on disputes between States can only be alleviated by the thought that the members of the Council are presumed to be well versed in matters of international civil aviation and therefore would be deemed to be better equipped to comprehend the issues placed before them than the distinguished members of the International Court of Justice, some of whom may not be experts of international air law. Nonetheless, there is no doubt that the ICAO Council possess juridical powers[54] and that as one commentator said:

> If ICAO did not exist, it would have to be invented; otherwise, international civil aviation would not function with the safety, efficiency and regularity that it has achieved today.[55]

International organizations can generally only work on the basis of legal powers that are attributed to them. Presumably, these powers emanate from the sovereign States that form the membership of such organizations. Therefore, the logical conclusion is that if international organizations were to act beyond the powers accorded to them, they would be presumed to act *ultra vires* or beyond the scope of their mandate. The universal solidarity of UN Contracting States that was

---

[50] Rules for the Settlement of Differences, ICAO Doc 7782/2 (2ed. 1975).

[51] *Id.* Article 14(a).

[52] Buergenthal, infra note 66, 121 at p. 136.

[53] Michael Milde, Dispute Settlement in the Framework of the International Civil Aviation Organization (ICAO), *Settlement of Space Law Disputes* (1979) 87, at p. 88.

[54] J.C. Sampayo de Lacerda, A Study About the Decisions of the ICAO Council...111 *Annals of Air and Space Law* (1978) at p. 219.

[55] Gerald F. Fitzgerald, ICAO Now and in the Coming Decades, *International Air Transport Law, Organization and Politics for the Future* (N.M. Matte ed., 1976) 47 at p. 50.

recognized from the outset at the establishment of the Organization brings to bear the need for States to be united in recognizing the effect of UN policy and decisions. This principle was given legal legitimacy in the 1971 decision concerning the European Road Transport Agreement handed down by the Court of Justice of the European Community. The court held that the competence of the European Community to conclude an agreement on road transport could not be impugned since the member States had recognized Community solidarity and that the Treaty of Rome which governed the Community admitted of a common policy on road transport which the Community regulated.

It should be noted that the United Nations does not only derive implied authority from its Contracting States based on universality but it also has attribution from States to exercise certain powers. The doctrine of attribution of powers comes directly from the will of the founders, and in the UN's case, powers were attributed to the Organization when it was established as an international organization that would administer the provisions of the United Nations Charter. In addition, the UN could lay claims to what are now called "inherent powers" which give it power to perform all acts that the Organization needs to perform to attain its aims not due to any specific source of organizational power but simply because the United Nations inheres in organizationhood. Therefore, as long as acts are not prohibited in the UN's constituent document (the Un Charter), they must be considered legally valid.

Over the past two decades the inherent powers doctrine has been attributed to the United Nations Organization and its specialized agencies on the basis that such organizations could be stultified if they were to be bogged down in a quagmire of interpretation and judicial determination in the exercise of their duties. The advantages of the inherent powers doctrine is twofold. Firstly, inherent powers are functional and help the organization concerned to reach its aims without being tied by legal niceties. Secondly, it relieves the organization of legal controls that might otherwise effectively preclude that organization from achieving its aims and objectives. The ability to exercise its inherent powers has enabled the UN to address issues on promoting self determination and independence, strengthening international law, handing down judicial settlements of major international disputes, providing humanitarian aid to victims of conflict, alleviating chronic hunger and rural poverty in developing countries and promoting women's rights, just to name a few.

6.1.4 Is ICAO a Judicial Body?

The words "legislative power" have been legally defined as "power to prescribe rules of civil conduct",[56] while identifying law as a "rule of civil conduct". The word "quasi" is essentially a term that makes a resemblance to another and classifies it. It is suggestive of comparative analogy and is accepted as:

---

[56] *Schaake* v. *Dolly* 85 Kan. 590., 118 Pac. 80.

the conception to which it serves as an index and its connection with the conception with which the comparison is instituted by strong superficial analogy or resemblance.[57]

The question *stricto sensu* according to the above definition is therefore whether the ICAO Council now has power to prescribe rules of civil conduct (legislative power) or in the least a power that resembles by analogy the ability to prescribe rules of conduct (quasi-legislative power). Since legislative power is usually attributed to a State, it would be prudent to inquire, on a general basis, whether the ICAO Council has law making powers (in a quasi-legislative sense). Therefore, all references hereafter that may refer to legislative powers would be reflective of the Council's law making powers in a quasi-legislative sense.

Article 54(l) of the Convention on International Civil Aviation prescribes the adoption of international Standards and Recommended Practices (hereafter, SAR-PS) and their designation in Annexes to the Convention, while notifying all con-tracting States of the action taken. The adoption of SARPS was considered a priority by the ICAO Council in its Second Session (2 September-12 December 1947)[58] which attempted to obviate any delays to the adoption of SARPS on air navigation as required by the First Assembly of ICAO.[59] SARPS inevitably take two forms: a negative form e.g. that States shall not impose more than certain maximum requirements; and a positive form e.g. that States shall take certain steps as prescribed by the ICAO Annexes.[60]

Article 37 of the Convention obtains the undertaking of each contracting State to collaborate in securing the highest practical degree of uniformity in regulations, standards, procedures and organization in relation to international civil aviation in all matters in which such uniformity will facilitate and improve air navigation. Article 38 obligates all contracting States to the Convention to inform ICAO immediately if they are unable to comply with any such international standard or procedure and notify differences between their own practices and those prescribed by ICAO. In the case of amendments to international Standards, any State which does not make the appropriate amendment to its own regulations or practices shall give notice to the Council of ICAO within 60 days of the adoption of the said amendment to the international Standard or indicate the action which it proposes to take.

The element of compulsion that has been infused by the drafters of the Con-vention is compatible with the "power to prescribe rules of civil conduct" on a *stricto sensu* legal definition of the words "legislative power" as discussed above. There is no room for doubt that the 18 Annexes to the Convention or parts thereof lay down rules of conduct both directly and analogically. In fact, although there is a

---

[57] *People* **v.** *Bradley* 60 Ill. 402, at p. 405. Also, *Bouviers Law Dictionary and Concise Ency-clopedia* 3 ed. Vol. 11, Vernon Law Book Co., New York 1914.

[58] Proceedings of the Council 2nd Session 2 September–12 December 1947, Doc 7248—C/839 at pp. 44–45.

[59] ICAO Resolutions A-13 and A-33 which resolved that SARPS relating to the efficient and safe regulation of international air navigation be adopted.

[60] ICAO Annex 9, Facilitation, Ninth Edition, July 1990, Foreword.

conception based on a foundation of practicality that ICAO's international Standards that are identified by the words "contracting States shall" have a mandatory flavour (infused by the word "shall") while Recommended Practices identified by the words "contracting States may have only an advisory and" recommendatory connotation (infused by the word "may"), it is interesting that at least one ICAO document requires States under Article 38 of the Convention, to notify ICAO of all significant differences from both Standards and Recommended Practices, thus making all SARPS regulatory in nature.[61]

Another strong factor that reflects the overall ability and power of the Council to prescribe civil rules of conduct (and therefore legislate) on a strict interpretation of the word is that in Article 22 of the Convention each contracting State agrees to adopt all practical measures through the issuance of special regulations or otherwise, to facilitate and expedite air navigation... It is clear that this provision can be regarded as an incontrovertible rule of conduct that responds to the requirement in Article 54(l) of the Convention. Furthermore, the mandatory nature of Article 90 of the Convention—that an Annex or amendment thereto shall become effective within 3 months after it is submitted by the ICAO Council to contracting States is yet another pronouncement on the power of the Council to prescribe rules of State conduct in matters of international civil aviation. A fortiori, it is arguable that the Council is seen not only to possess the attribute of the term "jurisfaction" (the power to make rules of conduct) but also the term "jurisaction" (the power to enforce its own rules of conduct). The latter attribute can be seen where the the Convention obtains the undertaking of contracting States not to allow airlines to operate through their air space if the Council decides that the airline concerned is not conforming to a final decision rendered by the Council on a matter that concerns the operation of an international airline.[62] This is particularly applicable when such airline is found not to conform to the provisions of Annex 2 to the Convention that derives its validity from Article 12 of the Convention relating to rules of the air.[63] In fact, it is very relevant that Annex 2, the responsibility for the promulgation of which devolves upon the Council by virtue of Article 54(l), sets mandatory rules of the air, making the existence of the legislative powers of the Council an unequivocal and irrefutable fact.

Academic and professional opinion also favours the view that in a practical sense, the ICAO Council does have legislative powers. Michael Milde says:

---

[61] *Aeronautical Information Services Manual*, ICAO Doc 8126-0 AN/872/3. ICAO Resolution A 1-31 defines a Standard as any specification for physical characteristics...the uniform application of which is recognised as *necessary*... and one that States *will conform to*. The same resolution describes a Recommended Practice as any specification for physical characteristics... which is recognised as *desirable*... and one that member States *will endeavour to conform to*... Buergenthal infra note 280, at p. 10 also cites the definitions given in ICAO's Annex 9 of SARPS.

[62] Article 86 of the Convention.

[63] Article 12 stipulates that over the high seas, the rules in force shall be those established under the Convention, and each contracting State undertakes to insure the prosecution of all persons violating the applicable regulations.

The Chicago Convention, as any other legal instrument, provides only a general legal framework which is given true life only in the practical implementation of its provisions. Thus, for example, Article 37 of the Convention relating to the adoption of international standards and recommended procedures would be a very hollow and meaningless provision without active involvement of all contracting States, Panels, Regional and Divisional Meetings, deliberations in the Air Navigation Commission and final adoption of the standards by the Council. Similarly, provisions of Article 12 relating to the rules of the air applicable over the high seas, Articles 17 to 20 on the nationality of aircraft, Article 22 on facilitation, Article 26 on the investigation of accidents, etc., would be meaningless without appropriate implementation in the respective Annexes. On the same level is the provision of the last sentence of Article 77 relating to the determination by the Council in what manner the provisions of the Convention relating to nationality of aircraft shall apply to aircraft operated by international operating agencies.[64]

Professor Milde concludes that ICAO has regulatory and quasi-legislative functions in the technical field and plays a consultative and advisory role in the economic sphere.[65] A similar view had earlier been expressed by Buergenthal who states:

the manner in which the International Civil Aviation organization has exercised its regulatory functions in matters relating to the safety of international air navigation and the facilitation of international air transport provides a fascinating example of international law making... the Organization has consequently not had to contend with any of the post war ideological differences that have impeded international law making on politically sensitive issues.[66]

Paul Stephen Dempsey endorses in a somewhat conservative manner, the view that ICAO has the ability to make regulations when he states:

In addition to the comprehensive, but largely dormant adjudicative enforcement held by ICAO under Articles 84–88 of the Chicago Convention, the Agency also has a solid foundation for enhanced participation in economic regulatory aspects of international aviation in Article 44, as well as the Convention's Preamble.[67]

A significant attribute of the legislative capabilities of the ICAO Council is its ability to adopt technical standards as Annexes to the Convention without going through a lengthy process of ratification.[68] Eugene Sochor refers to the Council as a powerful and visible body in international aviation.[69] It is interesting however to note that although by definition, the ICAO Council has been considered by some as unable to deal with strictly legal matters, since other important matters come within its

---

[64] Michael Milde, The Chicago Convention—After Forty Years, 1X *Annals Air and Space L.* 119, at p. 126. See also Jacob Schenkman, *International Civil Aviation Organization*, op. Cit. at p. 163.

[65] Milde., *id.* 122.

[66] T. Buergenthal, Law Making in the International Civil Aviation Organization, Syracuse University Press: Syracuse, 1969 at p. 9.

[67] Paul Stephen Dempsey, *Law and Foreign Policy in International Aviation*, Transnational Publishers Inc., Dobbs Ferry New York 1987, at p. 302.

[68] Eugene Sochor, *The Politics of International Aviation*, McMillan, London, 1991, at p. 58.

[69] *Ibid.*

purview,[70] this does not derogate the compelling facts that reflect the distinct law making abilities of ICAO. Should this not be true, the functions that the Convention assigns to ICAO in Article 44—that ICAO's aims and objectives are to "develop the principles and techniques of international air navigation and to foster the planning and development of international air transport"—would be rendered destitute of effect.

Under the Interim Agreement[71] the PICAO Council was required to act as an arbitral body on any differences arising among member States relating to matters of international civil aviation which may be submitted to it, wherein the Interim Council of PICAO was empowered to render an advisory report or if the parties involved so wished, give a decision on the matter before it.[72] The Interim Council, which was the precursor to the ICAO Council, set the stage therefore for providing the Council with unusual arbitral powers which are not attributed to similar organs of the specialised agencies of the United Nations system.[73] A fortiori, since the ICAO Council is permanent and is almost in constant session, contracting States could expect any matter of dispute brought by them before the Council to be dealt with, without unreasonable delay.[74]

Chapter XV111 of the Convention formalises the arbitral powers of the Council by stating:

> If any disagreement between two or more contracting States relating to the interpretation or application of this Convention and its Annexes cannot be settled by negotiation, **it shall** (emphasis added), on the application of any State concerned in the disagreement, be decided by the Council...[75]

This provision reflects two significant points: the first is that contracting States should first attempt to resolve their disputes by themselves, through negotiation[76]; the second is that the word **shall** in this provision infuses into the decision making powers of the Council an unquestionably mandatory character. Furthermore, a decision taken by the Council is juridically dignified by Article 86 of the Convention, when the Article states that unless the Council decides otherwise, any decision by the Council on whether an international airline is operating in conformity with the provisions of the Convention shall remain in effect unless reversed

---

[70]   Alexander Tobolewski, ICAO's Legal Syndrome... 1 V *Annals Air and Space L.* 1979, 349 at p. 359.

[71]   See note 41.

[72]   Interim Agreement, *op.cit.*, Article 111, Section 6(i).

[73]   Schenkman, *op.cit.* 160.

[74]   See statement of R. Kidron, Israeli Head Delegate, Statement of the Second Plenary Meeting of the Seventh Assembly on June 17, 1953, reported in *ICAO Monthly Bulletin*, August–October 1953, at p. 8.

[75]   Article 84.

[76]   Hingorani, Dispute Settlement in International Civil Aviation 14 *Arb J* 14, at p. 16 (1959). See also, *Rules of Procedure for the Council*, Fifth Edition 1980, Article 14.

in appeal. The council also has powers of sanction granted by the Convention, if its decision is not adhered to.[77] Schenkman states:

> The power of sanctions in this field is an entirely new phenomenon, attributed to an aeronautical body... none of the pre-war instruments in the field of aviation had the power of sanctions as a means of enforcement of its decisions.[78]

Most contracting States have, on their own initiative, enacted dispute-settlement clauses in their bilateral air services agreements wherein provision is usually made to refer inter-State disputes relating to international civil aviation to the ICAO Council, in accordance with Chapter XV111 of the Convention. In this context, it is also relevant to note that the President of the Council is empowered by the Convention to appoint an arbitrator and an umpire in certain circumstances leading to an appeal from a decision of the Council.[79]

A most interesting aspect of the ICAO Council remains to be that one of its mandatory functions is to consider any subject referred to it by a contracting State for its consideration.[80] or any subject which the President of the Council or the Secretary General of the ICAO Secretariat desires to bring before the Council.[81] Although the Council is bound to consider a matter submitted to it by a contracting State it can refrain from giving a decision as the Council is only obligated to consider a matter before it.

There seems to be an unfortunate dichotomy in terminology in the Convention since on the one hand, Article 54(n) makes it mandatory that the Council shall merely consider any matter relating to the Convention which any contracting State refers to it, while on the other, Article 84 categorically states that any disagreement between two or more States relating to the interpretation or application of the Convention and its Annexes, that cannot be settled by negotiation shall... be decided by the Council. The difficulty arises on a strict interpretation of Article 54 (n) where even a disagreement between two States as envisaged under Article 84 could well be considered as 'any matter' under Article 54(n). In such an instance, the Council could well be faced with the dilemma of choosing between the two provisions. It would not be incorrect for the Council to merely consider a matter placed before it, although a decision is requested by the applicant State, since, Article 54(n) is perceived to be comprehensive as the operative and controlling

---

[77] Article 87.

[78] Schenkman, *op.cit.* 162.

[79] Article 85.

[80] Rules of Procedure for the Council, *op.cit.* Section 1 V, Rule 24 (e). Also, Article 54(n) stipulates that one of the mandatory functions of the Council is to consider any matter relating to the Convention which any contracting State refers to it.

[81] Rules of Procedure for the Council, *op.cit.* Section 1 V Rule 24 (f). The two additional multilateral agreements stemming from the Convention and providing for the exchange of traffic rights—the Air Services Transit Agreement and the Air Transport Agreement, also contain provisions that empower the ICAO Council to hear disputes and "make appropriate findings and recommendations..." see Air Services Transit Agreement Article 11 Section 1, and the Air Transport Agreement Article 1 V Section 2.

provision that lays down mandatory functions of the Council. It is indeed unfortunate that these two provisions obfuscate the issue which otherwise would have given a clear picture of the decision making powers of the Council. A further thread in the fabric of adjudicatory powers of the Council is found in Article 14 of the Rules of Settlement promulgated by the Council in 1957[82] which allows the Council to request the parties in dispute to engage in direct negotiations at any time.[83] This emphasis on conciliation has prompted the view that the Council, under Article 84 would favour the settling of disputes rather than adjudicating them.[84] This view seems compatible with the proposition that the consideration of a matter under Article 54(n) would be a more attractive approach *in limine* in a matter of dispute between two States.

Dempsey points out that in the four decades since the promulgation of Chapter XV111, only three disputes had been submitted to the Council for formal resolution[85]: the first involved a dispute between India and Pakistan (1952), where India complained that Pakistan was in breach of the Convention by not permitting Indian aircraft to overfly Pakistani airspace on their way to Afghanistan; the second was a complaint filed by the United Kingdom against Spain (1969), alleging the violation by Spain of the Convention by the establishment of a prohibited zone over Gibraltar; and the third was a complaint by Pakistan against India (1971), concerning a hijacking of Indian aircraft which landed in Pakistan. India unilaterally suspended Pakistan's overflying privileges, five days after the hijacking. The first complaint was amicably resolved by the parties, the second was differed by the parties *sine die*, and the third was suspended with the formation of the State of Bangladesh in 1972, even though the matter had been processed as an appeal from the Council to the International Court of Justice. Unfortunately, none of these instances was taken to its conclusion so that the World could have had the opportunity to evaluate clearly, ICAO's decision making process.

Milde noted in 1979:

> The Council of ICAO cannot be considered a suitable body for adjudication in the proper sense of the word - i.e. settlement of disputes by judges and solely on the basis of respect for law. The Council is composed of States (not independent individuals) and its decisions would always be based on policy and equity considerations rather than on pure legal grounds...truly legal disputes...can be settled only by a true judicial body which can bring into the procedure full judicial detachment, independence and expertise. The under-employed ICJ is the most suitable body for such types of disputes.[86]

---

[82] Rules for the Settlement of Differences, ICAO Doc 7782/2 (2ed. 1975).

[83] *Id.* Article 14(a).

[84] Buergenthal, *op.cit.* note 66 at p. 136.

[85] Paul Stephen Dempsey, *op.cit.* at p. 295.

[86] Michael Milde, Dispute Settlement in the Framework of the International Civil Aviation Organization (ICAO), *Settlement of Space Law Disputes* (1979) 87, at p. 88.

The perceived inadequacies of the ICAO Council in being ethically unsuitable to decide on disputes between States can only be alleviated by the thought that the members of the Council are presumed to be well versed in matters of international civil aviation and therefore would be deemed to be better equipped to comprehend the issues placed before them than the distinguished members of the International Court of Justice, some of whom may not be experts of international air law. Nonetheless, there is no doubt that the ICAO Council possess juridical powers[87] and that as one commentator said:

> If ICAO did not exist, it would have to be invented; otherwise, international civil aviation would not function with the safety, efficiency and regularity that it has achieved today.[88]

International organizations can generally only work on the basis of legal powers that are attributed to them. Presumably, these powers emanate from the sovereign States that form the membership of such organizations. Therefore, the logical conclusion is that if international organizations were to act beyond the powers accorded to them, they would be presumed to act *ultra vires* or beyond the scope of their mandate. The universal solidarity of UN Contracting States that was recognized from the outset at the establishment of the Organization brings to bear the need for States to be united in recognizing the effect of UN policy and decisions. This principle was given legal legitimacy in the 1971 decision concerning the European Road Transport Agreement handed down by the Court of Justice of the European Community. The court held that the competence of the European Community to conclude an agreement on road transport could not be impugned since the member States had recognized Community solidarity and that the Treaty of Rome which governed the Community admitted of a common policy on road transport which the Community regulated.

6.1.5 Nuclear Power and Commercial Space Transportation

The International Atomic Energy Agency, in 1994 adopted the *Convention on Nuclear Safety*[89] which places responsibility for nuclear safety with the State having jurisdiction over a nuclear installation and obliges States Parties to take legislative and administrative action to implement the obligations imposed by the Convention.

---

[87] J.C. Sampayo de Lacerda, A Study About the Decisions of the ICAO Council…111 *Annals of Air and Space Law* (1978) at p. 219.

[88] Gerald F. Fitzgerald, ICAO Now and in the Coming Decades, *International Air Transport Law, Organization and Politics for the Future* (N.M. Matte ed., 1976) 47 at p. 50.

[89] The Convention on Nuclear Safety was adopted on 17 June 1994 by a Diplomatic Conference convened by the International Atomic Energy Agency at its Headquarters from 14 to 17 June 1994. See International Atomic Energy Agency, *INFCIRC/449 5 July 1994.*

In 1963, the United States, the United Kingdom and the then USSR signed in Moscow the *Treaty Banning Nuclear Weapons Tests in the Atmosphere, in Outer Space and Under Water,* commonly called the *Limited Test Ban Treaty* [90] The Treaty, although adopted outside the formal United Nations umbrella, is open to adoption by all States of the world.

The *Limited Test Ban Treaty* in Article 1 provides that each State Party to the treaty undertakes to prohibit, to prevent, and not carry out any nuclear weapon test explosion at any place under its jurisdiction or control, whether it be in the atmosphere, beyond its limits including outer space or over the high seas.[91]

With regard to outer space, the Treaty provides in Article 1 that States Parties to the Treaty undertake to prohibit, to prevent and not carry out any nuclear weapon test explosion, or any other nuclear explosion, at any place under its jurisdiction or control or *inter alia* in outer space or in any other environment if such explosion causes radioactive debris to be present outside the territorial limits of the State under whose jurisdiction or control such explosion is conducted.[92] The Treaty further provides that no State Party may collaborate with or encourage such activity.[93]

The *Outer Space Treaty,*[94] while expostulating the fundamental principle in its Article 1 that the exploration and use of outer space, including the moon and other celestial bodies, shall be carried out for the benefit and in the interests of all countries, explicitly imposes in Article VII international liability and responsibility on each State Party to the Treaty, for damage caused to another State Party or to its populace (whether national or juridical) by the launch or procurement of launch or launch from its territory or facility of an object into outer space. In its Article VI the Treaty imposes international responsibility on States Parties for national activities conducted in outer space. However, there are no reports that the DPRK launch

---

[90] *Treaty Banning Nuclear Weapon Tests in the Atmosphere, in Outer Space and Over Water,* signed at Moscow on 5 August 1963. UN Treaty Series, Volume 480, United Nations: 1965 at p. 45. In the 1970s the Disarmament Commission of the United Nations served as forum to coerce the super powers to amend the Partial Test Ban Treaty and transform it into a Comprehensive Test Ban Treaty. It will be recalled that, over the past decade, DPRK has been the focus of the United Nations in the area of nuclear weapons and arms control. In 1992 DPRK concluded a safeguards agreement with the International Atomic Energy Agency (IAEA) where DPRK accepted six IAEA inspections before it refused IAEA access to its Yongbyon Nuclear Reactor Complex, giving as a reason the fact that the Complex was a conventional military facility. Consequently, after the matter was referred by the IAEA to the Security Council of the United Nations, DPRK gave formal notice in 1993 of withdrawal from the Nuclear Non Proliferation Treaty. See *United Nations Legal Order,* Volume 1, Oscar Schachter and Christopher C. Joyner ed., American Society of International Law and Cambridge University Press: 1995, at pp. 314–315.

[91] *Treaty Banning Nuclear Weapon Tests in the Atmosphere, in Outer Space and Under Water, Id.* Article 1.1(a).

[92] *Id.,* Article 1(a) and (b).

[93] *Id.,* Article 2.

[94] *Treaty on Principles Governing the Activities of States in the Exploration and Use of Outer Space, including the Moon and Other Celestial Bodies,* opened for signature at London, Moscow, and Washington, 27 January 1967, 610 UNTS 205.

caused any damage to other States or their citizens. The Treaty also requires its States Parties to be guided by the principle of co-operation and mutual assistance in the conduct of all their activities in outer space.[95] This overall principle is further elucidated in Article III which exhorts States Parties to the Treaty to carry on activities in the exploration and use of outer space, including the Moon and other celestial bodies in accordance with international law, including the Charter of the United Nations, in the interest of maintaining international peace and security and promoting international cooperation and understanding.

6.1.6 Diplomacy

Another strength in favour of ICAO is its diplomatic machine which could come in handy from a political standpoint, in the event of disputes on commercial space activities. What is becoming increasingly clear in the progression of air law is that there are two main issues concerning the evolving role of air law. The first is that that the distinction between air law and space law is continuing to blur. The diplomatic implications of this blend of two major areas lie in the need to mesh the two disciplines to form one area of activity. The second is that principles of air law are getting increasingly involved in activities related to military warfare. Air law and space law are closely inter-related in some areas and both these disciplines have to be viewed in the 21st Century within the changing face of international law and politics. Both air law and space law are disciplines that are grounded on principles of public international law, which is increasingly becoming different from what it was a few decades ago. We no longer think of this area of the law as a set of fixed rules, even if such rules have always been a snapshot of the law as it stands at a given moment.

The second issue—relating to the impact of military warfare on evolving principles of air law is not entirely new, although there has been a growing momentum in recent times. The use of SAMs and anti-tank rockets by terrorists goes back to 1973. On 5 September 1973 Italian police arrested five Middle-Eastern terrorists armed with SA-7s. The terrorists had rented an apartment under the flight path to Rome Fiumicino Airport and were planning to shoot down an El Al airliner coming in to land at the airport. This arrest proved a considerable embarrassment to Egypt because the SA-7s were later traced back to a batch supplied to it by the Russian

---

[95] *Id.* Article IX. The Space Liability Convention of 1972 elaborated the above liability provision in more detail, see *Convention on International Liability for Damage Caused by Outer Space Objects*, opened for signature at London, Moscow and Washington, 29 March 1972, 961 UNTS 187. The *Moon Agreement* of 1979 provides that in the exploration and use of the moon, States Parties shall take measures *inter alia* to avoid harmfully affecting the environment of the earth through the introduction of extra terrestrial matter or otherwise. See *Agreement Governing the Activities of States on the Moon and other Celestial Bodies*, opened for signature at New York on 18 December 1979, 1363 UNTS 3. UN Doc. A/RES/34/68 of 5 December 1079.

Union. It was alleged that the Egyptian government was supplying some of the missiles to the Libyan army but inexplicably, the SA-7s had been directly rerouted to the terrorists. This incident also placed the Russian Union in an awkward position because of the possibility that its new missile and its policy of the proxy use of surrogate warfare against democratic states were revealed to the West.

The consequences of the nuclear missile[96] firings of 5 July 2006 by the Democratic Peoples' Republic of Korea (DPRK)[97] brought to bear the hazards and grave dangers such activities pose to civil aviation. Consequently, the President of the Council of ICAO, by letter dated 6 July 2006 addressed to the Director General of the Civil Aviation Administration of DPRK drew attention to the fact that the preliminary investigation of the President indicated that missiles launched by DPRK crossed several international air routes over the high seas. It was further revealed that, when extrapolating the projected paths of some of the missiles, it appeared that they could have interfered with many more air routes, both over Japan and the air space of the North Pacific Ocean.

In his letter, the President of the ICAO Council reminded DPRK that, as a signatory to the Chicago Convention DPRK has assumed obligations to comply with provisions of the Convention and Annexes thereto.[98] Accordingly, the President requested information from DPRK relating to the missile firing, in order that he may determine any further action to be taken.[99]

This is not the first instance of its kind addressed by ICAO. At its 32nd Session in September/October 1998, the ICAO Assembly adopted Resolution A 32-6

---

[96] A nuclear missile is a weapon which derives its destructive force from nuclear reactions of either nuclear fission or the more powerful fusion. As a result, even a nuclear weapon with a relatively small yield is significantly more powerful than the largest conventional explosives, and a single weapon is capable of destroying an entire city.

[97] The missile firings of DPRK prompted the unanimous adoption by the United Nations Security Council on 15 July 2006 of a Resolution imposing sanctions on DPRK's dangerous weapons and condemning the missile tests. In response to a protest and consequent request by Japan of a resolution to be adopted, the Security Council discussed the missile firing for 10 days. The Resolution adopted requires all UN member States to stop imports and exports of any material or funds relating to the reclusive DPRK's missile programmes or weapons of mass destruction. It also demands that DPRK suspend all activities related to its ballistic missile programme and re-establish a moratorium on the launch of missiles.

[98] The provisions of the Chicago Convention, which is an international treaty, are binding on contracting States to the Convention and therefore are principles of public international law. The International Court of Justice (ICJ), in the *North Sea Continental Shelf Case* (*I.C.J. Report10*, at p. 32), held that legal principles that are incorporated in Treaties become customary international law by virtue of Article 38 of the 1969 Vienna Convention on the Law of Treaties. See *Vienna Convention on the Law of Treaties*, United Nations General Assembly Document *A/CONF.39/27*, 23 May 1969.

[99] The ICAO Council is known to take a serious view of acts of States that constitute violations endangering the safety of international civil aviation. For example, the destruction of Flight KE 007 on 31 August 1983 prompted the Council to launch a full inquiry at which the President of the Council observed: "It falls clearly to ICAO ... to focus its attention on gaining a full and complete technical understanding of how this tragic event occurred and to examine every element in ICAO's existing technical provisions for promoting the safety of air navigation." See C-MIN 137/15 contained in ICAO Doc C-MIN 137th Session, 1992 at p. 131.

relating to safety of air navigation which addressed a similar incident which took place on 31 August 1998 in the same vicinity in which the DPRK missiles were fired. The ICAO Assembly exhorted all Contracting States to strictly comply with the provisions of the Chicago Convention, its Annexes and related procedures[100] The 1998 Resolution was also sparked off by a protest by Japan of a violation of its airspace by DPRK's test of what was believed to be a ballistic missile. DPRKs response was that the event at issue was not a missile launch but rather the country's first launch of an artificial Earth satellite.[101]

Resolution A 32-6 involved an incident on August 31, 1998, where an object propelled by rockets was launched by a DPRK and a part of the object hit the sea in the Pacific Ocean off the coast of Sanriku in northeastern Japan. The impact area of the object was in the vicinity of the international airway A590 which is known as composing NOPAC Composite Route System, a trunk route connecting Asia and North America where some 180 flights of various countries fly every day. The ICAO Assembly did not have any difficulty in determining that the launching of such an object vehicle was done in a way not compatible with the fundamental principles, standards and recommended practices of the Chicago Convention. The Assembly therefore urged all Contracting States to reaffirm that air traffic safety is of paramount importance for the sound development of international civil aviation, and to strictly comply with the provisions of the Chicago Convention its Annexes and its related procedures, in order to prevent a recurrence of such potentially hazardous activities.[102]

Clearly the issue at stake is that DPRK's missiles, which were fired over the high seas, could have adversely affected civil aircraft plying air routes over sovereign territories which extended to the questioned area over the high seas in which the missiles were cruising. In this regard, the fundamental legal postulate regarding rights over the high seas is contained in Article 87 of the *United Nations Convention on the Law of the Sea* (UNCLOS)[103] which provides that the high seas is open to all States and that freedom of the high seas comprises *inter alia* freedom of over flight, which gives aircraft of States the right to use airspace over the high seas without hindrance. Article 88 of the Convention expressly stipulates that the high seas shall be reserved for peaceful purposes. Furthermore, Article 300 of UNCLOS requires State parties to fulfil in good faith its obligations assumed under the Convention in a manner which would not constitute an abuse of right.

---

[100] ICAO State Letter AN 13/29-98/95 of 16 December 1998.

[101] *Britannica Book of the Year 1999* (Events of 1998), Encyclopaedia Britannica Inc: Chicago 1998 at p. 41.

[102] With regard to a later incident, it was reported by South Korean media that on September 9, 2004 there had been a large explosion at the Chinese/North Korean border, in North Korea's second northernmost province of Ryanggang. This explosion left a crater visible by satellite and precipitated a large (2 mile diameter) mushroom cloud.

[103] The Law of the Sea, Official Text of the United Nations Convention on the Law of the Sea with Annexes and Indexes, United Nations: New York, 1983 at p. 31.

From an aeronautical perspective, Annex 11 to the Chicago Convention,[104] which deals with the subject of air traffic services, lays down requirements for coordination of activities that are potentially hazardous to civil aircraft. Standard 2.17.1 stipulates that arrangements for activities potentially hazardous to civil aircraft, whether over the territory of a State or over the high seas, shall be coordinated with the appropriate air traffic services authorities, such coordination to be effected early enough to permit timely promulgation of information regarding the activities in accordance with the provisions of Annex 15 to the Chicago Convention.[105] Standard 2.17.2 of Annex 11 explains that the objective of the coordination referred to in the earlier provision shall be to achieve the best arrangements that are calculated to avoid hazards to civil aircraft and minimize interference with the normal operations of aircraft.

In Article 3 *bis* of the Chicago Convention, Contracting States *inter alia* recognize that every State must refrain from resorting to the use of weapons against civil aircraft in flight. In an instance where, as observed by the President of the ICAO Council, missiles fired by one State could endanger the aircraft of another, it can be argued that DPRK knew or ought to have known that its activity could well have resulted in damage envisioned by Article 83 *bis* and therefore DPRK should have given due notice to all concerned a required by Annex 11 to the Chicago Convention. The ICJ in the *Barcelona Traction Case* held:

[A]n essential distinction should be drawn between the obligations of a State towards the international community as a whole, and those arising *vis a vis* another State in the field of diplomatic protection. By their very nature, the former are the concerns of all States. In view of the importance of the rights involved, all States can be held to have a legal interest in their protection; they are obligations *erga omnes*.[106]

The International Law Commission (ILC)[107] has observed of the ICJ decision:

[I]n the Courts view, there are in fact a number, albeit limited, of international obligations which, by reason of their importance to the international community as a whole, are- unlike others - obligations in respect of which all States have legal interest.[108]

The ICAO Council has the authority, under Article 54(J) of the Chicago Convention, to report to Contracting States any infraction of the Convention (and

---

[104] Air traffic Services: Annex 11 to the Convention on International Civil Aviation, Thirteenth Edition, July 2001.

[105] Annex 15 contains Standards and Recommended Practices relating to Aeronautical Information Services.

[106] *Barcelona Traction, Light and Power Company Limited, I.C.J. Reports, 1974*, 253 at pp. 269–270.

[107] The International Law Commission was established by the General Assembly of the United Nations in 1947 with the declared objective of promoting the progressive development of international law and its codification. It consists of 34 members from Asia, Africa, The Americas and Europe. Most international treaties and conventions have been the result of the Commission's work. See generally, B. Ramcharan, *The International Law Commission*, Leiden: 1977.

[108] *Yearbook of International Law Commission* 1976, Vol. II, Part One at p. 29.

its Annexes since Annexes to the Convention are an integral part of the Convention) and also, under Article 54(k) to report to the Assembly any infraction of the Convention where a Contracting State has failed to take appropriate action within a reasonable time after notice of the infraction. This would of course mean that if there is no timely response from DPRK to the request of the President of the Council for information relating to the firing of the missiles, the Council is empowered to take action within the scope of Article 54 of the Chicago Convention.

Annexes 11 and 15 to the Chicago Convention both derive their legal legitimacy from the *Preamble* to the Convention which sets the tone and philosophy of the Convention. The *Preamble* states *inter alia* that the future development of international civil aviation can greatly help to create and preserve friendship and understanding in the world and any abuse of that development could become a threat to general security. This statement echoes the basic premise of Article 1, Paras 1 and 2 of the United Nations Charter which conveys that the primary purpose of the United Nations is to maintain international peace and security and to that end to take effective collective measures for the prevention and removal of threats to the peace.

A powerful message ICAO can bring with it to the regulation of commercial space transport is that When dealing with issues of aviation to which diplomacy is applied, it is important to remember that from the distant past, it has been recognized that a nation's air power is the sum total of all its civil and military aviation resources.[109] Furthermore, the importance of aviation toward maintaining peace has been accepted since World War 2 and is aptly reflected in the Statement of the British at that time, that civil aviation holds the key to power and importance of a nation and therefore it must be regulated or controlled by international authority.[110] Lord Beaverbrook for the British Government of that time stated in Parliament:

> Our first concern will be to gain general acceptance of certain broad principles whereby civil aviation can be made into a benign influence for welding the nations of the world together into a closer cooperation…it will be our aim to make civil aviation a guarantee of international solidarity, a mainstay of world peace.[111]

The intensely political overtones that moulded the incipient civil aviation system of the world immediately after the War, thereby incontrovertibly establishing the relevance of diplomacy, international politics and international relations in civil aviation, is borne out by the statement of the first President of the ICAO Council when he said:

---

[109] Parker van Zandt, *Civil Aviation and Peace*, Washington, 1944 at pp. 28, 93.

[110] Wings for Peace—Labour's Post War Policy for Civil Flying, published by the Labour Party of England, April 1944, cited in van Zandt, *supra* note 372 at p. 1.

[111] *Flight*, Vol. XLV No. 1331, January 27, 1944, at pp. 97–98.

> It is well that we should be reminded…if the extent of the part which diplomatic and military considerations have played in international air transport, even in periods of undisturbed peace. We shall have a false idea of air transport's history, and a very false view of the problems of planning its future, if we think of it purely as a commercial enterprise, or neglect the extent to which political considerations have been controlling in shaping its course.[112]

In retrospect, it must be noted that this statement is a true reflection of what civil aviation stood for at that time, and, more importantly, that the statement has weathered the passage of time and is true even in the present context. A more recent commentator correctly observes that over the past decades, civil aviation has had to serve the political and economic interests of States and that, in this regard, ICAO has alternated between two positions, in its unobtrusive diplomatic role and its more pronounced regulatory role.[113]

An inherent characteristic of aviation is its ability to forge inroads into human affairs and promote international discourse. It also promotes international goodwill and develops "a feeling of brotherhood among the peoples of the world". [114] Therefore, it has been claimed that problems of international civil aviation constitute an integral part of the universal political problems of world organization and therefore aviation problems cannot be solved without involving the world political and diplomatic machinery.[115] It is at these crossroads that one encounters the profound involvement of the United Nations mechanism in general and ICAO in particular, while extending the message of air transport to commercial space transport.

6.2   Direction

If ICAO were to take on new responsibilities in the commercial space transportation area, and in order to serve its 191 member States, irrespective of whether they are in the developed or developing category, ICAO has to justify its performance and values based stature. In other words, ICAO needs to undergo a whole system change. For ICAO's Business Plan (triennial budget which also serves as the Business Plan) to be implemented and results to be produced, firstly ICAO's leaders (its Council and the senior managers of the Secretariat) may have to drive the process of transition from service to performance both in air and space issues. They need to be the ambassadors of the Organization's mission and vision statements and set values and behaviours. They must "talk the talk" and "walk the walk".

---

[112] See Warner 1942, p. V.

[113] See Sochor 1991, p. xvi.

[114] See Schenkman 1955b, p. 6.

[115] *Id.* vi.

The next step is to ensure that the mission and vision statement influences all decision making. This should permeate right to the bottom of the ICAO Secretariat. Thirdly, the new culture and its results must be measured by causal performance indicators. In other words, ICAO's new culture should be constantly monitored. The final measure would be to ensure that the values of the Organization's culture pervade and drive every aspect of decision making and be seen in every system and process.

In a way, ICAO is already undergoing a cultural transformation. It has come a long way in developing a mission and vision statement and a business plan driven by strategic objectives. There is a leadership that is committed to its work but it lacks direction, purpose and above all, prioritization of work. For example, ICAO's legal work is not focused, often resulting in ineptitude. This will not bode well if ICAO were to take on an involvement in commercial space transport in addition to its overall role in international civil aviation. ICAO needs strong tools and aggressive goals have to be in place through a robust and energized operational plan that is not disaggregated among the Organization's constituent bureaux and other offices.

Such an operational plan must have objectives and key performance indicators, as in the end it is measurement that matters. For this there must be targets set, not just improvement of performance. In this regard it must be noted that ICAO's Business plan is on the right track, as it has all three types of indicators: *causal indicators*—which relate to values and behaviours (which are known in other words as core competencies); *output indicators*—which measure performance in terms if efficiency and productivity; and *outcome indicators*, which relate to the result or effect on clients and stakeholders.

Of these indicators, ICAO's concentration should be mainly on output indicators that measure productivity, efficiency, quality, innovation, creativity of the Organization as a whole and ensuing customer satisfaction. Innovation and creativity are key factors that serve to promote ICAO's contribution to its member States. Just as an example, since many States do not have the volume and scope of aviation activities which generate the resources and the base-line activity necessary to support a workable safety oversight system, ICAO's role must be to take the leadership in providing States with templates of different models of safety oversight and recommend what is best suited for them. ICAO could also further the involvement of regional safety oversight organizations that are successful; and provide guidance to States as to the modalities of the transfer of responsibilities or tasks, depending on the model used, from participating State to regional safety oversight organizations.

Leadership is the key to ICAO's role in the Twenty First Century, and nowhere is its need more pronounced than in the economic field. The inhibitive mind-set created by the Chicago Convention where Article 44 ascribes to ICAO only a watered down role to merely "foster the planning and development of international air transport" would militate against any cultural transformation that would demand a greater leadership role from ICAO. The archaic and hopelessly obsolete premise that ICAO's aim should be "to meet the needs of the peoples of the world for safe, regular, efficient and economical air transport" is diametrically opposed to the new thinking in the ICAO leadership that, in a competitive world, ICAO has to perform

100 % in safety and security in helping develop an air transport system that is sustainable and efficient. Therefore, ICAO needs a mindset change if it were to take on economic aspects of commercial space transportation, as its Secretary General indicates in his letter to Commander Hadfield.

ICAO's leadership role in the economic and technical fields of civil aviation hinges on two key factors: an aggressive operational plan with key performance indicators and targets; and the realization that organizational culture, which is an intangible asset, is the new frontier of competitive advantage. The latter is particularly important under the current circumstances of ICAO where human resources and expertise are in short supply. Cultural transformation starts with the leadership and individual and leadership values. When one looks at ICAO's current leadership structure, there is no room for doubt that this is not in short supply. However the trick is to motivate the staff sufficiently so that they would be impelled to follow their leaders in the transformation and forge ICAO's leadership forward in its various areas of work.

All this leads one to the bottom line, which is the need for change in the mindset of the Organization, from its service role to a role of implementation and assistance. The human factor is an essential consideration in this metamorphosis. The key and the starting point, however, is to recognize the need for the transition, which ICAO has already done. The next step is to recognize that ICAO needs its peoples' best efforts, both individually and collectively. ICAO's image and the perception of the outside world of ICAO as an effective Organization is anchored on the extent to which its workers represent themselves as good stewards of ICAO's business. They should therefore work together in the overarching interest of the Organization.

References

Abeyratne RIR (2005) The E-passport and the public key directory—consequences for ICAO. Air and Space Law XXX(4–5):255–268

Crawford J (2002) The international law commission's articles on state responsibility: introduction, text and commentaries. Butterworth, London, p 77

de Witte B (1998) Sovereignty and European integration: the weight of tradition. In: Slaughter A-M et al (eds) The European Court and National Courts: doctrine and jurisprudence. Hart Publishing, Oxford, pp 277–304

Klabbers J (2002) An introduction to international institutional law. Cambridge University Press, Cambridge, p 60

MacKenzie D (2008) ICAO, a history of the international civil aviation organization. University of Toronto Press, Toronto

Sarooshi D (2005) International organizations and their exercise of sovereign powers. Oxford University Press, Oxford, p 110

Schenkman J (1955a) International civil aviation organization. Librairie E. Droz, Geneve, p 125

Schenkman J (1955b) International civil aviation organization. Librairie E. Droz, Geneve, p 6

Seyersted F (1963) Objective international personality of intergovernmental organizations: do their capacities really depend upon the conventions establishing them? Copenhagen, p 28

Sochor E (1991) The politics of international aviation. Macmillan, London, p xvi

Warner E (1942) Foreword to international air transport and national policy. Lissitzyn O.J., New York, p V

Chapter 7
Conclusion

Based on the doctrine of empowerment, ICAO's member States can invoke their authority to empower ICAO to take on commercial space transportation activities as well, in accordance with a legal process. It cannot happen overnight and, as some commentators have claimed, happen by ICAO's unilateral change of its "mandate" or scope of functions. This book has outlines both the legal process that might apply, either through an amendment to the Chicago Convention itself i.e. by including Articles *bis* to the Convention in accordance with Article 94 of the Convention, or by adopting a separate multilateral treaty that would empower ICAO to address and involve itself in commercial air transportation as the designated specialized agency of the United Nations for that purpose. The latter is seemingly the cleaner, clearer and less messy approach.

Either of these measures would involve a substantial outlay of resources which member States would have to absorb, particularly since ICAO does not have any staff member in its Secretariat (not to mention the Council, its subsidiary bodies and in particular the current membership of the Legal Committee which has, on several occasions demonstrated its ineptitude in air law issues, assisted by an equally inept Legal Affairs and External Relations Bureau) who is qualified, knowledgeable and competent in commercial air transport issues.

It would particularly be foolish to have technical staff (which includes the likes of former air traffic controllers, pilots and radio engineers, who undoubtedly have legitimate claims to their own professions) go blindly into an exercise that would just arbitrarily and capriciously change or adapt existing provisions of the Chicago Convention and its Annexes. ICAO would first need a solid base document drawn in accordance with ICAO's legal status and applicable principles of modern treaty law and practice.

With regard to the Annexes to the Chicago Convention (if an additional Annex on commercial space transport is added, by whatever procedural pyrotechnics ICAO chooses) it must be mentioned that in this equation, the role of ICAO becomes an important one. Over its 70 years of service to the international civil aviation community, ICAO has, through its Assembly and Council adopted numerous Resolutions. Additionally, the ICAO Council has taken several decisions and issued statements of policy guidance. An organization such as ICAO is tasked

© The Author(s) 2015

R. Abeyratne, *Regulation of Commercial Space Transport*, SpringerBriefs in Law, DOI 10.1007/978-3-319-12925-9_7

primarily to provide a certain predictability about its members by promulgating norms for the conduct of its Member States. Of course not all those norms are binding and not all of them are adopted with the same degree of formality. However, certainly all of them provide guidance to States. This situation has to mesh with the basic inquiry as to whether ICAO, as an international organization, has been given direct authority over individuals or States. Another issue is whether ICAO is primarily an instrument for cooperation among States.

Firstly, when one considers the background of ICAO and the statements of its founding fathers, and as discussed earlier, there is no room for doubt that ICAO is a specialized agency that has procedures to modify, without eliminating, the positivist principle that States are only bound by international rules to which they have consented. This approach admits of a process whereby ICAO adopts or amends rules after having given a designated period of time for its member States to examine such rules and decide whether they would accept them or not. Individual member States may object or mark their differences in practices to the ones ICAO suggests for adoption.[1] States objecting to a particular Standard and Recommended Practice (SARP) may choose if they wish to opt out of whole processes recommended by ICAO, even though general consensus is achieved to adopt them. There is no record of a single international Standard adopted through this process being disapproved by a majority of ICAO member States, although not all of ICAO's 190 member States have found it practicable to comply with all Standards[2] in the 18 Annexes to the Chicago Convention.[3]

The question arises as to whether a member State is formally bound by Standards contained in an Annex to the Chicago Convention, particularly when such a State has no convincing argument that it is impracticable to implement such Standards or when it has not notified the ICAO Council of differences as required. This is a vexed debate, particularly in the face of two blatant facts. The first is that the *travaux préparatoires* to the Convention contains a statement that "the Annexes

---

[1] Article 37 of the Chicago Convention confirms that each Member State undertakes to collaborate in securing the highest practicable degree of uniformity in regulations, standards, procedures and organization in relation to aircraft, personnel, airways and auxiliary services in all matters in which such uniformity will facilitate and improve air navigation. Article 38 gives any State the opportunity, if it finds it impracticable to adhere to ICAO's policy to file differences by giving notice to ICAO of the difference between what is recommended or required by ICAO and the practice prevalent in that State.

[2] The ICAO Assembly, at its 35th Session held in Montreal from 28 September to 8 October 2004, defined a Standard "as any specification...the uniform application of which is recognized as necessary for the safety or regularity of international air navigation and to which member States will conform in accordance with the Chicago Convention.; in the event of impossibility of compliance, notification to the Council is compulsory under Article 38 of the Convention. The same resolution describes a Recommended Practice as any specification for physical characteristics... which is recognised as desirable and one that member States will endeavour to conform to" See *Assembly Resolutions in Force*, (As of 8 October 2004) ICAO Doc 9848, II-2 Appendix A.

[3] Buergenthal, *Law Making in The International Civil Aviation Organization*, Syracuse, pp. 98–107.

are given no compulsory force".[4] The second is that in Article 54 of the Convention, which lays down the mandatory functions of the Council, it is provided that one of the mandatory functions is to

> Adopt, in accordance with the provisions of this Convention, international standards and recommended practices; **for convenience** (emphasis added) designate them as Annexes to this Convention; and notify all member States of action taken.[5]

One could argue therefore that the Annexes are not an integral part of the Convention by virtue of the statement in Article 54 and therefore do not form binding law.

There have been numerous views of legal scholars who have cautioned against this approach and advocated that the words of the Convention should not be taken literarily. One commentator is of the view that:

> The debate is largely academic. Whether or not ICAO standards are formally binding in the treaty law sense, they are highly authoritative in practice. This reflects their recognized importance for the safety and efficiency of civil air travel and the thorough process by which they are promulgated.[6]

7.1 Procedures for Enforcing SARPs

This is a purely procedural provision and the historical and legal bases of the Annexes have already been discussed under Articles 37 and 38. The ICAO Assembly, at its 2nd Session (Geneva 1–21 June 1948) adopted Recommendation 8 which recommended that the Council, when adopting further Annexes and establishing a date by which States may notify their disapproval of them, take fully into account the time needed for transmitting the Annexes, so as to allow for their effective study during the full period provided in Article 90. In Resolution A7-9 (adopted at the 7th Session of the Assembly—Brighton 16 June–6 July 1953) the assembly resolved that the Council, in fixing the dates for the application by

---

[4] See Whiteman 1968, p. 404.

[5] Chicago Convention, *supra*, Article 54(l).

[6] Frederic L. Kirgis, Jr, Specialized Law Making Processes, *United Nations Legal Order*, Volume 1 Chapter Two, Oscar Schachter and Christopher Joiner ed., The American Society of International Law: 1995, pp. 109 at 126. There is a similar process in operation under the World Meteorological Organization, whereby a certain amount of decision making authority is given to the WMO Congress. Article 9(a) of the WMO Convention provides that all members shall do their utmost to implement the decisions of the Congress. Article 9(b) allows any member to opt out by notifying the Secretary General, with reasons if it finds it impracticable to give effect to the technical requirement in question. WMO Convention, reprinted in *International Organization and Integration*, (P. Kapteyn et al. eds) 2nd Revised Edition, 1981, pt. I.B.1.9 a. Also in *WMO Basic Documents*, No. 1. WMO Doc. No. 15 at p. 9, 1987.

Contracting States of International Standards, allow sufficient time to enable States to complete their arrangements for implementation thereof.

There are two operative phrases of importance in Article 90 which speak of an Annex being "effective" within 3 months after its submission to the Contracting States and the Annex "coming into force". The issue is: "what is the significance of these two practices?"

7.2 Difference Between Standards and Recommended Practices

Firstly one must start with the two main components of the Annexes i.e. Standards on the one hand, and Recommended Practices on the other. A Standard is defined as any specification for physical characteristics, configuration, material, performance, personnel or procedure, the uniform application of which is recognized as necessary for the safety or regularity of international air navigation and to which Contracting States will conform in accordance with the Convention; in the event of impossibility of compliance, notification to the Council is compulsory under Article 38 of the Convention.

A Recommended Practice is any specification for physical characteristics, configuration, material, performance, personnel or procedure, the uniform application of which is recognized as desirable in the interest of safety, regularity or efficiency of international air navigation, and to which Contracting States will endeavour to conform in accordance with the Convention. States are invited to inform the Council of non-compliance.

SARPs are formulated in broad terms and restricted to essential requirements. For complex systems such as communications equipment, SARPs material is constructed in two sections: core SARPs—material of a fundamental regulatory nature contained within the main body of the Annexes, and detailed technical specifications placed either in the Appendices to Annexes or in manuals.

The differences to SARPS notified by States are published in Supplements to Annexes. After the Council adopts an Annex it is sent to ICAO member States for their comments and notification of disapproval of Standards with a date identified by the Council of the Annexes effective date which is within 3 months of submission. This is an interim edition of the Annex, referred to as the "Green Edition", which is dispatched to States with a covering explanatory letter. This covering letter also gives the various dates associated with the introduction of the Annex including its effective date. Once the Annex becomes effective, the States have 3 months to indicate disapproval of adopted amendments to SARPs. Unless a majority of the States indicate their disapproval on or before the time allocated to them to respond with their disapprovals, the Council declares the Annex to have come into force at a particular date.

All this leaves one with the inevitable question as to whether ICAO has sufficient clout to enforce its mandatory duties which appear under Article 54 of the Chicago Convention and in particular Article 54j). From a legal perspective, the above discussion points to a resolute "yes" with an additional qualifier that it is indeed ICAO's duty to do so. It is therefore largely left to ICAO to decide on the path it takes to ensure the legitimacy of its SARPs and the credibility of its Assembly and Council in the most diplomatic manner possible, in this defining point in the history of the Organization.

Finally, it must be mentioned that the instrument reflecting ICAO's involvement with commercial space transport, regardless in which that instrument comes into effect—whether by the amendment provision in the Chicago Convention or by a separate multilateral treaty, will be binding only on those member States that accept it. responsibility could devolve upon ICAO with regard to commercial space transport regulation by an amendment to the convention or a new treaty, either through a merger with another organization or through the transfer of functions to a specific branch or bureau of ICAO such as the Air Navigation Bureau.[7]

If a separate treaty is drawn up, which would be the most uncomplicated approach, there are certain facts to be borne in mind.[8] The Vienna Convention on the Law of Treaties, provides that such treaties shall be binding upon the Parties and be performed by them in good faith. The Vienna Convention further states that a treaty shall be interpreted in good faith in accordance with the ordinary meaning to be given to the terms of the treaty. Furthermore, the Vienna Convention stipulates that, unless a different intention appears from the treaty or is otherwise established, a treaty is binding upon each Party in respect of its entire territory. Therefore, a part of a State, however formed as a province, cannot take a unilateral decision to contravene the provisions of a treaty which the sovereign State has entered into. The Convention goes on to state that a Party may not invoke the provisions of its internal law as justification for its failure to perform a treaty. A treaty is an international agreement concluded between States[9] in written form and governed by international law, whether embodied in a single instrument or in two or more related

---

[7] See Amerasinghe 2003, p. 473.

[8] ICAO is the depository of 35 international treaties and related instruments entered into between its member States, both on a bilateral and multilateral basis. In addition, ICAO performs registration functions with regard to other aeronautical agreements between States *inter se* and international organizations. These functions are provided by the Legal Affairs and External Relations Bureau. Many aeronautical treaties have emanated from initiatives of the ICAO Legal Committee. ICAO has a significant role to play in the procedural aspects of treaty work which brings to bear the need to discuss the salient principles of law and practice relating to aeronautical treaties and agreements. Such a study becomes essential to the understanding of treaties related to aviation security, safety and future agreements concerning aviation and environmental protection.

[9] A State has been defined in Article 1 of the Montevideo Convention of 1933 as having the following characteristics: a permanent population; a defined geographic territory; a government; and the legal capacity to enter into relations with other States. See Montevideo Convention on the Rights and Duties of States, Signed at Montevideo, 26 December 1933. The Convention entered into Force, 26 December 1934. At http://www.taiwandocuments.org/montevideo01.htm.

instruments and whatever its particular designation.[10] The above notwithstanding, a treaty can be concluded between a State and another subject of international law such as an international Organization. An example is the Headquarters Agreement between ICAO and the Government of Canada.[11] When a State places its signature on a treaty it merely means that the State has agreed to the text in the instrument. It comes into effect for that State when it is ratified[12] by the State. At the time of ratification a State can record a reservation to a part of the treaty.[13] These generic principles and those discussed below also apply to aeronautical treaties and agreements.

It must be noted that a State can sign a treaty in two ways. The first is called attestation by "simple signature" which corresponds to the above statement—that such a signature merely denotes that a State agrees with the text of an instrument and a simple signature is subject to ratification, acceptance or approval. However, if a State attaches to the instrument what is called a "definitive signature" it means that the State has agreed to be bound by the treaty. Therefore a definitive signature obviates the need for that State to later ratify the treaty, as it has the same force as ratification.

The process of ratification usually goes through two phases. The first is the internal procedure where the State concerned has to attend to its constitutional provisions by sending the text of the instrument it has signed through its national legislature or parliament. Once parliament adopts the text as its internal law, the State then has to proceed with its international procedure of depositing its notice of ratification with the depository. In formal terminology this process is called the doctrine of incorporation where customary international law as incorporated in a treaty that has been signed by a State is recognized as the internal law of the land on

---

[10] Vienna Convention on the Law of Treaties, 1969, Done at Vienna on 23c May 1969, United Nations General Assembly Document *A/CONF.39/27*, 23 May 1969, Article 2(a). The Convention entered into force on 27 January 1980. UNTS Vol. 1155, p. 331.

[11] Headquarters Agreement between Canada and ICAO of 14 April 1951, which paraphrased the 1947 Convention on the Privileges and Immunities of the Specialized Agencies. On 20 February 1992, the 1951 Agreement was terminated and superseded by a new Agreement that entered into force the same day. A new Supplementary Agreement was signed on 28 May 1999 superseding the Supplementary Agreement signed in 1980 in order to reflect the relocation of the Organization's Headquarters to a new location on 999 University Street on November 1, 1996. See Supplementary Agreement Between the International Civil Aviation Organization and the Government of Canada Regarding the Headquarters of the International Civil Aviation Organization, Doc 9591.

[12] "Ratification", "acceptance", "approval" and "accession" mean in each case the international act so named whereby a State establishes on the international plane its consent to be bound by a treaty. See Vienna Convention, *supra* note 388 at Article 2(b).

[13] "Reservation" means a unilateral statement, however phrased or named, made by a State, when signing, ratifying, accepting, approving or acceding to a treaty, whereby it purports to exclude or to modify the legal effect of certain provisions of the treaty in their application to that State. Id. 2(d).

the common law practice based on a presumption that the legislature does not intend to commit a breach of international law.[14]

As the Vienna Convention[15] provides, a treaty need not be signed. According to Article 12 of the Convention the consent of a State to be bound by a treaty is expressed by the signature of the representative of that State only in certain circumstances.[16] Article 13 goes on to say that the consent of States to be bound by a treaty constituted by instruments exchanged between them is expressed by that exchange when: the instruments provide that their exchange shall have that effect; or it is otherwise established that those States were agreed that the exchange of instruments should have that effect. States may also contract with each other under their domestic laws.

A treaty enters into force when the number of ratifications as specified in that treaty is received by the depository. When a treaty enters into force it is in force for only those States who have consented to be bound by it which are called "Parties".[17] However, an expression by a State that it consents to be bound by a particular treaty does not mean that *ipso facto* that treaty enters into force for that State. Either, the treaty must already be in force at that time, or as already mentioned the number of ratifications must be deposited. The Vienna Convention (1969) is more specific when it says that a treaty enters into force in such manner and upon such date as it may provide or as the negotiating States may agree.[18] There are three ways in which a treaty may enter into force. They are: on a date specified in the treaty; on signature only, as agreed by the negotiating States; or on ratification by all or a specified number of States. A treaty may be considered to apply to a State provisionally when the treaty itself so provides; or the negotiating States have in some other manner so agreed. Unless the treaty otherwise provides or the negotiating States have otherwise agreed, the provisional application of a treaty or a part of a treaty with respect to a State will be considered as terminated if that

---

[14] See *West Rand Central Gold Mining Company* v. *R.* [1905] 2. K.B. 391 per Lord Alverstone who stated in his judgment: "whatever has received the common consent of civilized nations must have received the assent of our country, and that to which we assented along with other nations in general may properly be called international law: and as such will be acknowledged and applied by our municipal tribunals to decide questions to which doctrines of international law may be relevant". *Id.* 397.

[15] *Supra* note 10.

[16] Article 12(1) states that the consent of a State to be bound by a treaty is expressed by the signature of its representative when: (a) the treaty provides that signature shall have that effect; (b) it is otherwise established that the negotiating States were agreed that signature should have that effect; or (c) the intention of the State to give that effect to the signature appears from the full powers of its representative or was expressed during the negotiation. Article 12.2 provides that for the purposes of para 1: (a) the initialling of a text constitutes a signature of the treaty when it is established that the negotiating States so agreed; (b) the signature ad referendum of a treaty by a representative, if confirmed by his State, constitutes a full signature of the treaty.

[17] Vienna Convention, *supra* note 10 at Article 2(1)(g). It should be noted that such States should not be called "signatories" as some refer to them erroneously.

[18] *Id.* Article 24(1).

State notifies the other States between which the treaty is being applied provisionally of its intention not to become a party to the treaty.[19]

Treaties, conventions, agreements, protocols, exchanges of notes and other synonyms all mean one and the same thing at international law—that they are international transactions of a legal character. Treaties are concluded between States in written form and governed by international law, whether embodied in a single instrument or in two or more related instruments and whatever its particular designation.[20] Each treaty has four constituent elements: the capacity of the parties thereto to conclude agreement of the provisions of the treaty under international law; the intention of the parties to apply principles of international law when concluding agreement under a treaty; *consensus ad idem* or a meeting of the minds of the parties[21]; and, the parties must have the intention to create legal obligations among themselves. These four elements form a composite regulatory process whereby a treaty becomes strong enough at international law to enable parties to settle their differences within the parameters of the treaty, make inroads into customary international law if necessary, and, transform an unorganized international community into one which may be organized under a uniform set of rules. Treaties are based on three fundamental principles of international law: good faith; consent; and fundamental international responsibility.[22] Since international customary law does not prescribe any particular form for consensual agreements and requirements that would make them binding, the parties to a treaty could agree upon the form of treaty they intend entering into and make it binding among them accordingly. Legal bonds are established between nations because they wish to create them and, as is seen in the *Preamble* to the Chicago Convention, a statement to this effect is reflected in the treaty itself.[23] The main feature of a multilateral international agreement is that absolute rights that may have existed within States before the entry into force of such treaty would be transformed into relative rights in the course of a balancing process in which considerations of good faith and reasonableness play a prominent part. However, treaty provisions must be so written and

---

[19] *Id*. Article 25.

[20] *Id*. Article 2(1)(a).

[21] There are instances where States may record their reservation on particular provisions of a convention while signing the document as a whole. The International Court of Justice in its examination of the *Genocide Convention* has ruled:

The object and purpose of the Convention... limit both the freedom of making reservations and that of objecting to them. It follows that it is the compatibility of a reservation with the object and purpose of the Convention that must furnish the criterion for the attitude of a State in objecting to the reservation. 1 *I.C.J Rep.* 1951, at p. 15.

[22] See Schwarzenberger and Brown 1976, p. 118.

[23] The *Preamble* to the Chicago Convention states:

... the undersigned governments having agreed on certain principles and arrangements in order that international civil aviation may be developed in a safe and orderly manner and that international air transport services may be established on the basis of equality of opportunity and soundly and economically; have accordingly concluded this Convention to that end.

construed as best to conform to accepted principles of international customary law.[24]

Great reliance is placed on treaties as a source of international law. The international Court of Justice, whose function it is to adjudicate upon disputes of an international character between States, applies as a source of law, international conventions which establish rules that are expressly recognized by the States involved in a dispute.[25] The Court also has jurisdiction to interpret a treaty at the request of a State.[26]

The Vienna Convention[27] while recognizing treaties as a source of law, accepts free consent, good faith and the *pacta sunt servanda* as universally recognized elements of a treaty.[28] Article 11 of the Vienna Convention provides that the consent of a State to be bound by a treaty may be expressed by signature, exchange of instruments constituting a treaty, ratification, acceptance, approval or accession, or by any other means agreed upon. "Ratification", "acceptance", "approval", and "accession" generally mean the same thing, i.e. that in each case the international act so named indicates that the State performing such act is establishing on the international plane its consent to be bound by a treaty. A State demonstrates its adherence to a treaty by means of the *pacta sunt servanda*, whereby Article 26 of the Vienna Convention reflects the fact that every treaty in force is binding upon the parties and must be performed by them in good faith. The validity of a treaty or of the consent of a State to be bound by a treaty may be impeached only through the application of the Vienna Convention[29] which generally requires that a treaty could be derogated upon only in circumstances the treaty in question so specifies[30]; a later treaty abrogates the treaty in question[31]; there is a breach of the treaty[32]; a *novus actus interveniens* or supervening act which makes the performance of the treaty impossible[33]; and the invocation by a State of the *Clausula Rebus Sic Stantibus*[34] wherein a fundamental change of circumstances (when such circumstances constituted an essential basis of the consent of the parties to be bound by the treaty) which has occurred with regard to those existing at the time of the conclusion of the treaty, and which was not foreseen by the parties, radically changes or transforms the extent of obligations of a State. A State may not invoke the fact that its consent

---

[24] See Greig 1976, p. 8.

[25] *Statute of the International Court of Justice, Charter of the United Nations and Statute of the International Court of Justice*, United Nations: New York, Article 38. 1(a).

[26] *Id.* Article 36.2(a).

[27] *Supra* note 10.

[28] *Vienna Convention*, Preamble and Article 26.

[29] *Id.* Article 42. 1.

[30] *Id.* Article 57.

[31] *Id.* Article 59.

[32] *Id.* Article 60.

[33] *Id.* Article 61.

[34] *Id.* Article 62.

to be bound by a treaty has been expressed in violation of a provision of its internal law regarding competence to conclude treaties and seek to invalidate its consent unless such violation was manifest and concerned a rule of its internal law of fundamental importance.[35]

States or international organizations which are parties to such treaties have to apply the treaties they have signed and therefore have to interpret them. Although the conclusion of a treaty is generally governed by international customary law to accord with accepted rules and practices of national constitutional law of the signatory States, the application of treaties is governed by principles of international law. If however, the application or performance of a requirement in an international treaty poses problems to a State, the constitutional law of that State would be applied by courts of that State to settle the problem. Although Article 27 of the Vienna Convention requires States not to invoke provisions of their internal laws as justification for failure to comply with the provisions of a treaty, States are free to choose the means of implementation they see fit according to their traditions and political organization.[36] The overriding rule is that treaties are juristic acts and have to be performed.

Every international treaty is affected by the fundamental dichotomy where on the one hand, the question arises whether provisions of a treaty are enforceable at law, and on the other, whether the principles of State sovereignty, which is *jus cogens* or mandatory law, would pre-emt the provisions of a treaty from being considered by States as enforceable. Article 53 of the Vienna Convention addresses this question and provides that where treaties, which at the time of their conclusion conflict with a peremptory norm of general international law or *jus cogens* are void. A peremptory norm of general international law is a norm accepted and recognized by the international community of States as a whole as a norm from which no derogation is permitted and which can be modified only by a subsequent norm of general international law having the same character. The use of the words "as a whole" Article 53 effectively precludes individual States from considering on a subjective basis, particular norms as acceptable to the international community.[37] According to this provision therefore, a treaty such as the Chicago Convention could not have derogated from principles of accepted international legal norms when it was being concluded. The Vienna Convention has, by this provision, implicitly ensured the legal legitimacy of international treaties, and established the principle that treaties are in fact *jus cogens* and therefore are instruments containing provisions, the compliance with which is mandatory.

Once a State ratifies a treaty it has to deposit instruments of ratification with the United Nations or with the specialized agency of the United Nations as prescribed

---

[35] *Id.* Article 46.

[36] See Reuter 1989, p. 16.

[37] See Frans G. von der Dunk, Jus Cogens Sive Lex Ferenda: Jus Cogendum, *Air and Space Law: De Lege Ferenda*, Essays in Honour of Henri A Wassenbergh, Tanja L. Masson-Zwaan and Pablo M.J. Mendes De Leon ed. Martinus Nijhoff: Dordrecht 1992, 219 at pp. 223–224.

in the treaty concerned. Such Instruments must emanate from and be signed by the Head of State, Head of Government or Minister for Foreign Affairs or a person exercising, *ad interim*, the powers of one of the above authorities; clearly identify the treaty concerned and the type of action consistent with the provisions of the treaty, i.e. ratification, acceptance, approval, accession, consent to be bound, etc.; contain an unambiguous expression of the will of the Government, acting on behalf of the State, to recognize itself as being bound by the treaty concerned and to undertake faithfully to observe and implement its provisions (a simple reference to a domestic statutory provision will be inadequate); indicate the title of the signatory. In the case of a person exercising, *ad interim*, the powers of the Head of State, Head of Government or Minister for Foreign Affairs, the title must indicate that the person is exercising such powers *ad interim*. In this respect, the depositary accepts the following formulations: Acting Head of State, Acting Head of Government, Acting Minister for Foreign Affairs, Head of State *ad interim*, Head of Government ad interim and Minister for Foreign Affairs ad interim; indicate the date and place where the instrument was issued;

If required, the instrument of deposit must specify the scope of its application in accordance with the provisions of the relevant treaty; and contain all mandatory declarations and notifications in accordance with the provisions of the relevant treaty. Where reservations are intended, such reservations since reservations must be signed by the Head of State, Head of Government or Minister for Foreign Affairs or a person exercising, ad interim, the powers of one of the above authorities.

A State is expected to be bound by a treaty it signs and later ratifies by the legal maxim *pacta sunt servanda* invoked in Article 27 of the Vienna Convention which provides that every treaty in force is binding upon the parties to it and must be performed by them in good faith. A State Party to a treaty can opt out of this obligation in two instances, the first being the recording of what is called "*an interpretative declaration*"[38] during the diplomatic conference that discusses the text of a Treaty in the context of a multilateral treaty or during bilateral discussions in the instance of negotiating a bilateral treaty with another State. These declarations are widely used and go back in history to 1815.[39] In essence such a declaration is made when a State has a difference of view with regard to the meaning of a particular provision. In such an instance the State concerned makes a formal statement expressing the interpretation favoured by it and that statement is usually reflected in the *Travaux Préaparatiores* or negotiating history (record of proceedings). It is not uncommon for a State to put forward an interpretative declaration even after a Treaty has been concluded and this occurrence is seen mainly in instances where a State realizes subsequently that a provision of a treaty it has

---

[38] An interpretative declaration is "a unilateral declaration, however phrased or named, made by a State or by an international Organization whereby that State or that Organization purports to clarify the meaning or scope attributed by the declarant to the treaty or to certain of its provisions". See UN Doc, A/CN.4/491/Add4, para. 361.

[39] Interpretative Declaration of the United Kingdom in respect of an instrument adopted during the Congress of Vienna, 1815. See 64 CTS 454.

signed and/or ratified is inconsistent or contrary to its domestic law in whole or in part thereof. A watered down version of an interpretative declaration is a *political declaration* which does not *per se* address the legality of a treaty provision in the eyes of a State but rather clarifies the State's policy towards that provision. By making such a declaration a State may keep open a window of opportunity that would enable the State to make a reservation at the point of ratification.

The second instance wherein a State could opt out of its obligation from adhering to an entire treaty is when it records a *reservation* to any particular provision in the Treaty at the point of definitive signature or ratification. Article 19 of the Vienna Convention provides that a State may, when signing, ratifying, accepting, approving or acceding to a treaty,[40] formulate a reservation unless: the reservation is prohibited by the treaty; the treaty provides that only specified reservations, which do not include the reservation in question, may be made; or in cases not failing under the abovementioned conditions, the reservation is incompatible with the object and purpose of the treaty. A reservation is:

> a unilateral statement, however phrased or named, made by a State, when signing, ratifying, accepting, approving or acceding to a treaty, whereby it purports to exclude or to modify the legal effect of certain provisions of the treaty in their application to that State.[41]

A reservation, an express acceptance of a reservation and an objection to a reservation must be formulated in writing and communicated to the contracting States and other States entitled to become parties to the treaty. If formulated when signing the treaty subject to ratification, acceptance or approval, a reservation must be formally confirmed by the reserving State when expressing its consent to be bound by the treaty. In such a case the reservation will be considered as having been made on the date of its confirmation. An express acceptance of, or an objection to a reservation made previously to confirmation of the reservation does not itself require confirmation.[42]

In practice a reservation need not necessarily be unilateral and two or more States can put forward the same reservation.[43] There is also a *derogation* which should be distinguished from a *reservation,* the former being a concession accorded by a treaty to States' Parties to derogate from a provision or provisions of that treaty to accommodate special exigencies such as a state of emergency.

---

[40] The Vienna Convention does not contain provision for States Parties to make reservations to a treaty subsequent to their ratifying a treaty. However, The Secretary General of the United Nations may circulate a reservation received subsequently with a note that, unless he hears otherwise from other States' Parties any objections to the reservation within 90 days the reservation will deem to have been accepted. This same practice may be applied when a State wishes to modify a reservation previously made. See Anthony Aust, *Modern Treaty Law and Practice* Cambridge University Press: 2011 at p. 129.

[41] Vienna Convention, *supra* note 10, Article 2(1)(d).

[42] *Id.* Article 23.

[43] Anthony Aust, supra note 418, at p. 263.

A reservation established with regard to another party modifies for the reserving State in its relations with that other party the provisions of the treaty to which the reservation relates to the extent of the reservation; and modifies those provisions to the same extent for that other party in its relations with the reserving State. The reservation does not modify the provisions of the treaty for the other parties to the treaty *inter se*. When a State objecting to a reservation has not opposed the entry into force of the treaty between itself and the reserving State, the provisions to which the reservation relates do not apply as between the two States to the extent of the reservation.[44]

A reservation can be withdrawn at any stage and such an instance usually occurs when the situation prevailing in a State at the time the reservation is made ceases to exist. The Vienna Convention expressly provides that unless the treaty otherwise provides, a reservation may be withdrawn at any time and the consent of a State which has accepted the reservation is not required for its withdrawal.[45] The withdrawal of a reservation or of an objection to a reservation must be formulated in writing.[46]

The *pacta sunt servanda* element of a treaty (consent to be bound) may be adversely affected in instances of State succession when one State succeeds wholly or in part to the legal personality of another State, both of whom are parties to the same treaty. Instances of war and armed conflict, although they do not automatically sever treaty relationships between States may affect them. A treaty is void if its conclusion has been procured by the threat or use of force in violation of the principles of international law embodied in the Charter of the United Nations.[47] The basic principle, following the Charter of the United Nations is that treaties are no less binding in instances of war. However, termination of a treaty may be by consent of the parties, express or implied. Article 54 of the Vienna Convention prescribes that the termination of a treaty or the withdrawal of a party may take place either in conformity with the provisions of the treaty; or at any time by consent of all the parties after consultation with the other contracting States. A treaty does not terminate merely because the number of Parties to that treaty falls below the required number to enter into force. The termination of a treaty under its provisions or in accordance with the Convention releases the parties from any obligation further to perform the treaty but does not affect any right, obligation or legal situation of the parties created through the execution of the treaty prior to its termination.[48]

As mentioned at the outset of this book, it is not prudent to mesh air transport and space transport by arbitrarily adding provisions simplistically to the Chicago Convention, nor is it wise to add on an Annex on commercial space transport to the

---

[44] Vienna Convention, *supra* note 10, Article 21.

[45] *Id.* Article 22.

[46] Id. Article 23(4).

[47] *Id.* Article 52.

[48] *Id.* Article 70.

Convention, however expeditious that might be as a quick fix. At the least, any involvement of ICAO should be separate from its responsibilities pertaining to civil aviation under the Chicago Convention. Of course, as already mentioned, a separate Convention involving ICAO would essentially involve an expansion and change in the scope, functions and title of ICAO which would have to be changed to reflect reality. Most importantly, additional scope and functions of ICAO would inevitably require additional resources, particularly in the form of experts in the legal and technical issues of space transport. There should be no political favouritism and nepotism just to get people into the organization at any cost, which would cost the international community dearly. Existing personnel at ICAO should not be given the task of handling space transport issues for which they have no expertise. ICAO has already bungled a few recent treaties with useless unqualified people at the helm, some of whom had no claims whatsoever to even a rudimentary knowledge of air law. There are lessons learnt.

The most important issue is that, before such an attempt the international space community should know precisely where they are headed in terms of the various definitions which seem to elude them.

References

Amerasinghe CF (2003) Principles of the institutional law of international organizations, 2nd edn. Cambridge Univesity Press, Cambridge, p 473
Greig DW (1976) International law, 2nd edn. Butterworths, London, p 8
Reuter P (1989) Introduction to the law of treaties. Pinter Publishers, London and New York, p 16
Schwarzenberger G, Brown ED (1976) A manual of international law, 6th edn. Professional Books Limited, Oxon, p 118
Whiteman M (1968) Digest of international law. Kluwer, The Netherlands, p 404